VOL. 3

Uncovering
Student Ideas
in Science

Another 25
Formative
Assessment
Probes

VOL. 3

Uncovering
Student Ideas
in Science

Another 25 Formative Assessment Probes

By Page Keeley,
Francis Eberle,
and Chad Dorsey

NSTApress
NATIONAL SCIENCE TEACHERS ASSOCIATION
Arlington, Virginia

National Science Teachers Association

Claire Reinburg, Director
Judy Cusick, Senior Editor
Andrew Cocke, Associate Editor
Betty Smith, Associate Editor

Cover, Inside Design, and Illustrations by Linda Olliver

Printing and Production
Catherine Lorrain, Director

National Science Teachers Association
Gerald F. Wheeler, Executive Director
David Beacom, Publisher

Library of Congress Cataloging-in-Publication Data

Keeley, Page.
Uncovering student ideas in science / by Page Keeley, Francis Eberle, and Lynn Farrin.
    v. cm.
Includes bibliographical references and index.
Contents: v. 1. 25 formative assessment probes
ISBN 0-87355-255-5
1. Science--Study and teaching. 2. Educational evaluation. I. Eberle, Francis. II. Farrin, Lynn.
III. Title.
Q181.K248 2005
507'.1--dc22
                          2005018770

NSTA is committed to publishing material that promotes the best in inquiry-based science education. However, conditions of actual use may vary, and the safety procedures and practices described in this book are intended to serve only as a guide. Additional precautionary measures may be required. NSTA and the authors do not warrant or represent that the procedures and practices in this book meet any safety code or standard of federal, state, or local regulations. NSTA and the authors disclaim any liability for personal injury or damage to property arising out of or relating to the use of this book, including any of the recommendations, instructions, or materials contained therein.

Contents

Preface

Introduction

Physical Science and Nature of Science Assessment Probes

Life, Earth, and Space Science Assessment Probes

Dedication

This book is dedicated to Dr. Gerry Wheeler, upon his retirement as executive director of the National Science Teachers Association. His leadership, vision, creativity, respect for science teachers, and commitment to all students will live on for many years.

Preface

Overview

This book is the third in the highly successful *Uncovering Student Ideas in Science* series. The addition of 25 more formative assessment probes has now expanded the collection to a total of 75 probes into student thinking in science—thinking that is rarely revealed through standard assessment questions. A new addition to the collection of Earth, space, physical, and life science probes is the inclusion of three probes that target the nature of science and science as inquiry. Together, the probes focus on important fundamental ideas in science that cut across multiple grade spans.

Regardless of whether you teach elementary, middle, or high school science, misconceptions are tenacious and often follow students from one grade to the next. Taking the time to elicit and examine student thinking is one of the most effective ways to support instruction that leads to conceptual change and enduring understanding. It is also the starting point for differentiating instruction to meet the content needs of all students.

Since Volume 1 was released in October 2005, thousands of teachers and hundreds of thousands of K–12 and university students have used the probes. The response has been very encouraging. Teachers have said that students actually ask for and look forward to the opportunity to use an assessment probe. Students eagerly ask teachers for "one of those probe things"—certainly not the typical student reaction when it comes to assessment!

Not only are teachers using probes to elicit students' ideas and inform instructional practices, but they have become a tool for transformative teacher learning. In our work at the Maine Mathematics and Science Alliance, we provide professional development to many school districts, math-science partnership projects, and other teacher enhancement initiatives throughout the United States that have embedded the use of these probes in their teacher professional development programs. Working with teachers has shown us that formative assessment is a powerful catalyst for engaging teachers in examining student learning and teacher practice. As a result of the growing interest in using these probes for teacher professional development, we decided to focus Volume 3 on considerations for using probes in a professional learning context.

While you are probably most interested in using the 25 probes in this book, don't overlook the Introduction (pp. 1–13) or the Introductions in Volumes 1 and 2. Each Introduction will expand your understanding of formative assessment and its inextricable link to instruction and learning. Volume 1 provides an overview of formative assessment, including what it is and how it differs from summative assessment. It also provides background on probes as specific types of formative assessments and

Preface

how they are developed. Volume 2 describes the link between formative assessment and instruction and suggests ways to embed the probes in your teaching. This volume (Volume 3) describes how you can use the probes and student work to deepen your understanding of teaching and learning.

Each probe is accompanied by an extensive Teacher Notes section that clarifies the probe and can be used to improve teachers' content knowledge of fundamental ideas in science as well as increase their knowledge of appropriate curricular emphasis and pedagogical implications. The Teacher Notes are made up of the following elements:

Purpose

This section describes the general concept or topic targeted by the probe and the specific idea being elicited by the probe. It is important to be clear about what the probe is going to reveal. Being clear will help you decide if the probe fits your intended target.

Related Concepts

Each probe is designed to target one or more related concepts that cut across grade spans. These concepts are described in the Teacher Notes and are also included on the matrix charts on pages 16 and 110. A single concept may be addressed by multiple probes. You may find it helpful to use a cluster of probes to target a concept or specific ideas within a concept. For example, there are three probes that target kinetic molecular theory.

Explanation

A brief scientific explanation, reviewed by scientists and content specialists, accompanies each probe and provides clarification of the scientific content that underlies the probe. The explanations are designed to help you identify what the best or most scientifically acceptable answers are (sometimes there is not a "right" answer) as well as clarify any misunderstandings you might have about the content. The explanations are not intended to provide detailed background knowledge on the concept, but enough to connect the idea in the probe with the scientific knowledge it is based on. If you have a need for further explanation of the content, several of the probe notes list NSTA resources, such as the series *Stop Faking It! Finally Understanding Science So You Can Teach It,* that will enhance and extend your understanding of the content.

Curricular and Instructional Considerations

The probes in this book are not limited to one grade level as summative assessments are. Rather, they provide insights into the knowledge and thinking that your students may have related to a topic as they progress from one grade span to the next. Some of the probes can be used for grades K–12, while others may cross over just a few grade levels. Teachers from two grade spans (e.g., middle school and high school) might decide to use the same probe and come together and discuss their findings. To do this, it is helpful to have insight into what students typically experience at a given grade span as it relates to the ideas elicited by the probe. Because the probes do not prescribe a specific grade level for use, you are encouraged to read the curricular and instructional considerations and decide if your students have had sufficient experience to make the probe useful.

The Teacher Notes also describe how the information gleaned from the probe is useful at a given grade span. For example, it might be useful for planning instruction when an idea in the probe is a grade-level expectation or it might be useful at a later grade to find out whether students have sufficient prior knowledge to move on to the next level. Sometimes the student learning data gained through use of the probe indicates that you have to back up several grade levels to teach ideas that are not really clear to students.

We deliberately chose not to suggest a grade level for each probe. If these were intended to be used for summative purposes, a grade level, aligned with a standard, would be suggested. However, these probes have a different purpose. Do you want to know about the ideas your students are expected to learn in your grade-level standards? Are you interested in how ideas develop and change across multiple grade levels in your school even when they are not formally taught? Are you interested in whether students have achieved a scientific understanding of previous grade level ideas before you introduce higher-level concepts? The descriptions of grade-level considerations in this section can be coupled with the section that lists related ideas in the national standards in order to make the best judgment about grade-level use.

Administering the Probe

In this section, we suggest ways to administer the probe, including a variety of modifications that may make the probe more useful at certain grade spans. For example, the Teacher Notes might recommend eliminating certain examples from a justified list for younger students who may not be familiar with particular words or examples, or using the word *weight* instead of *mass* with younger elementary students who might confuse the word *mass* with *massive*. The notes also include suggestions for demonstrating the probe context with artifacts and eliciting the probe responses while students interact within a group.

Related Ideas in the National Standards

This section lists the learning goals stated in the two national documents generally considered the "national standards": *Benchmarks for Science Literacy* (AAAS 1993) and *National Science Education Standards* (NRC 1996). Since the probes are not designed as summative assessments, the learning goals from those two documents are not intended to be considered alignments but rather as related ideas connected to the probe. Some targeted ideas, such as a student's conception of a life cycle on page 111, are not explicitly stated as learning goals in the standards but are clearly related to national standards concepts that address specific ideas about life cycles. When the ideas elicited by a probe appear to be a strong match with a national standard's learning goal, these matches are indicated by a star symbol (★).

Preface

Related Research

Each probe is informed by related research when it is available. Because the probes were not designed primarily for research purposes, an exhaustive literature search was not conducted as part of the development process. We drew primarily from two comprehensive research summaries commonly available to educators: Chapter 15 in *Benchmarks for Science Literacy* (AAAS 1993) and *Making Sense of Secondary Science: Research Into Children's Ideas* (Driver et al. 1994). Although both of these resources describe studies that have been conducted in past decades, and involved children not only in the United States but in other countries as well, many of the results of these studies are considered timeless and universal. Many of the ideas students held that were uncovered in the research during the 1980s and 1990s still apply today. It is important to recognize that cultural and societal contexts can influence students' thinking, but research also indicates that many of these ideas are pervasive regardless of geographic boundaries. Hence the descriptions from the research can help you better understand the intent of the probe and the variety of responses your students are likely to have. As you use the probes, you are encouraged to seek new and additional research findings. One source of updated research can be found on the Curriculum Topic Study (CTS) website at *www.curriculumtopicstudy.org*. A searchable database on this site links each of the CTS topics to additional research articles and resources.

Suggestions for Instruction and Assessment

After analyzing your students' responses, it is up to you to decide on appropriate student interventions and instructional planning. We have included suggestions gathered from the wisdom of teachers, the knowledge base on effective science teaching, and our own collective experience as former teachers and specialists involved in science education. These are not exhaustive or prescribed lists but rather suggestions that may help you modify your curriculum or instruction in order to help students learn ideas that they may be struggling with. It may be as simple as realizing that you need to provide a variety of contexts or that a specific strategy or activity might work with your students. Learning is a very complex process and most likely no single suggestion will help all students learn the science ideas. But that is part of what formative assessment encourages: thinking carefully about the variety of instructional strategies and experiences needed to help students learn scientific ideas.

Related NSTA Bookstore Publications and Journal Articles

NSTA's journals and books are increasingly targeting the ideas that students bring to their learning. We have provided suggestions for additional readings that complement or extend the use of the individual probes and the background information that accompanies them. For example, Bill Robertson's *Stop Faking It!* series may be helpful in clarifying content that teachers struggle with. A journal article from one of NSTA's elementary, middle school, or high school journals may provide additional insight into students' misconceptions or provide an example of an effective instructional strategy or activity that can be used to develop understanding of the ideas targeted by a probe. Other resources listed in this section provide a more comprehensive overview of the topic addressed by the probe.

Related Curriculum Topic Study Guides

NSTA is a copublisher of the book *Science Curriculum Topic Study: Bridging the Gap Between Standards and Practice* (Keeley 2005). This book was developed as a professional development resource for teachers with funding from the National Science Foundation's Teacher Professional Continuum Program. It provides a set of 147 curriculum topic study (CTS) guides that can be used to learn more about a science topic's content, examine instructional implications, identify specific learning goals and scientific ideas, examine the research on student learning, consider connections to other topics, examine the coherency of ideas that build over time, and link understandings to state and district standards. The CTS guides use national standards and research in a systematic process that deepens teachers' understanding of the topics they teach.

The probes in this book were developed using the CTS guides and the assessment tools and processes described in Chapter 4 of the CTS book. The CTS guides that were used to inform the development of each of the probes are listed. Teachers who wish to delve deeper into the standards and research-based findings that informed the development of the probe and are linked to its use in curriculum and instruction may wish to use the CTS guides for further information.

References

References are provided for the standards and research findings cited in the Teacher Notes.

As a companion to this book and the other two volumes, NSTA has copublished *Science Formative Assessment: 75 Practical Strategies for Linking Assessment, Instruction, and Learning* (Keeley 2008). In this book you will find a variety of strategies to use with the probes to facilitate elicitation, support metacognition, spark inquiry, encourage discussion, monitor progress toward conceptual change, encourage feedback, and promote self-assessment and reflection.

We hope this volume of probes will be as useful to you as the other two volumes. As the popular saying goes, "Build it and they will come," and indeed, you came. In concluding this third volume, we turn the cliché around to say, "Come and we will build it." In other words, we sincerely hope the demand for quality formative assessment continues—and, if it keeps coming, we will keep "building"!

Acknowledgments

The assessment probes in this book have been extensively field-tested by several teachers and hundreds of students in Maine, New Hampshire, Iowa, Missouri, Florida, and California by the Maine Mathematics and Science Alliance (MMSA). We would like to thank the teachers

Preface

and science coordinators we have worked with for their willingness to field-test probes, share student data, and contribute ideas for additional assessment probe development. In particular we would like to acknowledge the following people for their contributions to this volume: Dick Beer, Kennebunk, ME; Margaret Collins, Hooksett, NH; Bev Cox, Orlando, FL; Linda D'apolito, Falmouth, ME; Michelle DeBlois, Livermore Falls, ME; Kevin Fleury, Hooksett, NH; David Gerlach, Cedar Rapids, IA; Susan German, Hallsville, MO; Rita Harvey, Cathedral City, CA; Mark Koenig, Gardiner, ME; Kelle Kolkmeier, Cedar Rapids, IA; Elizabeth Lynch, Dover, NH; Anne MacDonald, Falmouth, ME; Molly Malloy, Orlando, FL; Kellie Martino, Hooksett, NH; Rainey Miller, Cedar Rapids, IA; Bonnie Mizzell, Orlando, FL; Margo Murphy, Thomaston, ME; Lisa Nash, Dover, NH; Betsy O'Day, Hallsville, MO; Michael Praschak, Lewiston, ME; Nicole Rodway, Hooksett, NH; Terri Schott, Cedar Rapids, IA; Amy Troiano, Poland, ME; Jane Voth-Palisi, Concord, NH; Joan Walker, Orlando, FL; Andy Weatherhead, Kittery, ME; Katherine Wheeler, Hillsboro, NH. We sincerely apologize if we overlooked anyone.

We would also like to thank our colleagues at the Maine Mathematics and Science Alliance *(www.mmsa.org)*, who continue to support us in this work, as well as our professional development colleagues, Math-Science Partnership Directors, and university partners throughout the United States, with whom we have had the pleasure of sharing this work. In addition, special thanks goes to Dr. Herman Weller at the University of Maine whose graduate seminar, attended by one of the authors, sparked an interest in the nature of science that subsequently went into developing the nature of science probes for this volume. And certainly, we are deeply appreciative of the efforts of the National Science Teachers Association (NSTA) in supporting formative assessment and the outstanding staff that make up the NSTA Press.

About the Authors

Page Keeley, senior science program director; Dr. Francis Eberle, executive director; and Chad Dorsey, science and technology specialist, all work at the Maine Mathematics and Science Alliance (MMSA) in Augusta, Maine, where they develop, support, and coordinate various science education initiatives throughout Maine and New England and provide professional development in formative assessment nationally. Combined, they have a total of over 30 years of teaching experience in middle and high school science and have served as adjunct instructors in the University of Maine system. They have worked with teachers, schools, and organizations in the areas of professional development, leadership, mentoring, standards, curriculum development, technology, assessment, and school reform. The authors have served as co-PIs and senior personnel on six National Science Foundation grants and three state Math-Science Partnership Projects. They currently serve on state and national advisory boards and committees and frequently present their work each year at the NSTA conferences.

In addition, Page Keeley is the 2008–2009 president of NSTA.

References

American Association for the Advancement of Science (AAAS). 1993. *Benchmarks for Science Literacy.* New York: Oxford University Press.

Driver, R., A. Squires, P. Rushworth, and V. Wood-Robinson. 1994. *Making sense of secondary science: Research into children's ideas.* London and New York: RoutledgeFalmer.

Keeley, P. 2005. *Science curriculum topic study: Bridging the gap between standards and practice.* Thousand Oaks, CA: Corwin Press.

Keeley, P. 2008. *Science formative assessment: 75 practical strategies for linking assessment, instruction, and learning.* Thousand Oaks, CA: Corwin Press.

National Research Council (NRC). 1996. *National science education standards.* Washington, DC: National Academy Press.

Introduction

When data are used by teachers to make decisions about next steps for a student or group of students, to plan instruction, and to improve their own practice, they help *inform* as well as *form* practice; this is *formative assessment.*

—Maura O'Brien Carlson, Gregg E. Humphrey, and Karen S. Reinhardt, *Weaving Science Learning and Continuous Assessment* (2003, p. 4.)

Imagine a team of fourth-grade teachers meeting after school with their district science coordinator to plan for implementation of a new curriculum unit, called "Silkworm." The unit, developed at a summer institute with the support of an entomologist university partner, addresses culminating learning goals related to the characteristics and needs of organisms, life cycles, and behaviors. The teachers are excited about using silkworms as a new context for learning about life cycles.

The unit is designed to help students understand that the life cycles of different organisms differ in their details, but all include a cycle of birth, growth and development, reproduction, and death.

The district science coordinator relates that in previous grades students investigated the life cycles of painted lady butterflies using one of the kit-based science programs: The students hatched frog eggs from the local pond, observed the development from tadpole to frog, and planted bean seeds to observe their development from seed to flowering plants. The teachers conclude that their fourth-grade students have sufficient prior knowledge of life cycles to begin the silkworm investigation without the need to review basic life-cycle concepts.

During the meeting, the district science coordinator alerts the teachers to a new book in their district resource library, *Uncovering Student Ideas in Science, Volume 3: Another 25 Formative Assessment Probes* (this book). One of the probes, shown in Figure 1, is titled, "Does It Have a Life Cycle?" The science coordinator shows the teachers the probe, along with the accompanying Teacher Notes. The notes indicate that some students tend to associate life cycles only with the examples they have encountered in school, such as certain types of plant, butterfly, frog, or mealworm life cycles or organisms that are similar to those they studied.

The teachers' curiosity is piqued by this finding. They wonder if their own students' understanding of life cycles might be limited by the examples and activities they have experienced in school. Are students able to make generalizations about life cycles beyond the individual organism they have studied? The teachers agree to give the probe to their students the next day and meet after school to look at the results.

The next day the teachers meet again and bring samples of their students' work.

Figure 1. "Does It Have a Life Cycle?" Probe

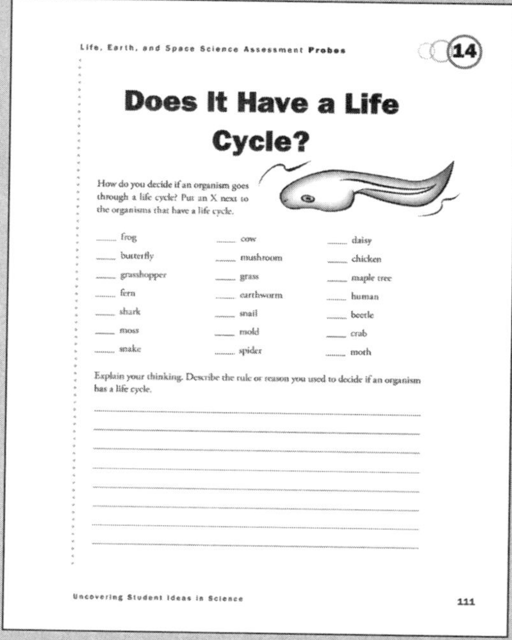
Life, Earth, and Space Science Assessment **Probes** 14

Does It Have a Life Cycle?

How do you decide if an organism goes through a life cycle? Put an X next to the organisms that have a life cycle.

____ frog	____ cow	____ daisy
____ butterfly	____ mushroom	____ chicken
____ grasshopper	____ grass	____ maple tree
____ fern	____ earthworm	____ human
____ shark	____ snail	____ beetle
____ moss	____ mold	____ crab
____ snake	____ spider	____ moth

Explain your thinking. Describe the rule or reason you used to decide if an organism has a life cycle.

Uncovering Student Ideas in Science 111

Their assumption that the students are ready to begin the unit with a firm foundation of basic life cycle ideas has been shattered: They noticed that the majority of students had little difficulty choosing "butterfly," "frog," and "chicken," and several chose "bean plant," yet they failed to check off other plants and animals. Few students checked off "human." The teachers think that the problem may be that their students have never explicitly encountered the bigger idea that every living plant and animal has a life cycle, even though that cycle may vary depending on the type of organism. Analyzing the students' reasoning, the teachers notice that several students hold similar context-bound ideas such as "it has to look very different at one time in its life," "it has to go through metamorphosis," and "it has to lay eggs." They even notice that a few students from each class correctly explained what a life cycle is yet failed to check off all the living things on the list.

As the teachers examine and discuss the data, pointing out similar misconceptions held by groups of students as well as a few idiosyncratic ideas, they realize how much they are learning about their students that they would not have known without using the probe. The district science coordinator asks the teachers to think about how they can use the data to modify the silkworm life cycle lessons so that their students can move past their context-bound ideas. The teachers discuss ways to challenge students' preconceptions, to explicitly connect the silkworm's life cycle to other organisms' life cycles, and to help students develop the broader generalization that all plants and animals go through a life cycle. The teachers develop a set of probing discussion questions for small groups that will challenge students' pre-formed ideas and lead to a whole-class discussion. They also decide to share their fourth-grade data with the first-, second-, and third-grade teachers.

The above example illustrates how formative assessment probes can uncover valuable information for teachers that often goes unnoticed in the science classroom and passes on from one grade to the next. The "Silkworm" unit is content rich and inquiry based, designed to develop students' understanding of life science concepts. It was designed to build on students' previous experiences with plant and animal life cycles, but it lacked a component that would uncover students' prior ideas about the basic concept of a life cycle. Students were asked to recall the details of the life cycles of the organisms they studied in previous years, yet they were not called upon to develop the fundamental idea that every plant and animal goes through a life cycle. It is likely that these students would have simply added the silkworm to their collection of "conceptions of life cycles by individual organisms"—rather than use a generalization that applies to all multicellular organisms they encounter in school and in everyday life—if their teachers had not taken the time to uncover their organism-specific concept of a life cycle. It is also likely that these teachers would not have uncovered this conception if they had not had formative assessment probes whose purpose was to reveal commonly held ideas.

Formative Assessment Probes

This book, as well as the two volumes that precede it, is designed to probe for commonly held ideas about fundamental concepts that

Introduction

can develop early in a student's education and persist all the way through high school if not identified and targeted for conceptual change instruction. This third collection of K–12 formative assessment probes continues to provide assessment examples that teachers can use to ask interesting questions, provoke lively discussions, encourage argumentation in small groups about differing ideas, orchestrate classroom discourse that promotes the public sharing of ideas, and examine students' ideas and reasoning through their written science explanations. These probes support students in being more metacognitive—that is, in becoming more aware of how and why they think about ideas in science—and also help teachers to aid individual students and the class progress toward developing scientific understanding (Keeley 2008).

Each assessment probe in this book, as well as those in Volumes 1 and 2, is a carefully designed question based on a formative assessment development process used in *Science Curriculum Topic Study* (Keeley 2005) that gives information to teachers about students' factual and conceptual understandings in science and the connections students make between and across ideas. Students respond to probes in writing as well as through small-group or class discussion, generating a range of ideas that help the teacher diagnose and address potential learning difficulties. Typically, teachers use the probes to identify potential misconceptions that can be barriers as well as springboards for learning, gather information on student thinking and learning in order to

make informed decisions to plan for or adjust instructional activities, monitor the pace of instruction, and spend more time on ideas that students struggle with. In addition to informing instruction and promoting student learning—purposes that are described extensively in Volumes 1 and 2 and in the complementary publication, *Science Formative Assessment: 75 Practical Strategies for Linking Assessment, Instruction, and Learning* (Keeley 2008)—assessment probes are powerful tools to enhance teacher learning.

Using Probes to Examine Teaching and Learning

Probes provide an entry point for teachers at all levels, from preservice to experienced teacher leaders, to examine and discuss the teaching and learning process, including the effective use of formative assessment. Some teachers think they are using formative assessment when they use probes to gather evidence of misconceptions but then proceed with their lesson plans despite their students' responses to the probes. There is little point in gathering data from formative assessment if the data are not used to fashion what comes next—"only when such refashioning occurs does the assessment become formative assessment" (Atkin and Coffey 2003, p. 6). Furthermore, the use of probes can extend beyond "refashioning" instruction. It can update the teachers' science content and pedagogical content knowledge and result in transformative teacher learning. Unlike additive learning, in which teachers acquire a new activity or strategy to add to their instructional repertoire, transformative learning results in changes in deeply held beliefs, knowledge, and habits of practice (Thompson and Zueli 1999).

Probes and Transformative Learning for Teachers

Using the probes as a transformative teacher learning experience fits with Thompson and Zueli's five requirements for transformative learning (Loucks-Horsley et al. 2003). A transformative learning experience should

1. *Create a high level of cognitive dissonance.* That is, new information, whether it is the content of the probe or the students' ideas as seen in students' written work or probe discussions, should reveal a "disconnect" between what the teacher thought he or she knew about the content (or about his or her students' ideas) and what is actually revealed through the probe. For example, a teacher might take the probe before using it with students and find that he or she has the same misconception that was revealed in the research. Or a high school teacher might believe that her chemistry students understand very basic matter concepts and be surprised to learn that her students have misconceptions about fundamental ideas.

2. *Provide sufficient time, structure, and support for teachers to think through the dissonance they experience.* Study groups, mentoring, coaching sessions, and professional learning communities give teachers a vehicle to make sense of the probe content, of the curricular and instructional implications of students' responses, and of their students' ways of reasoning.

3. *Embed the dissonance-creating and dissonance-resolving activities in teachers' actual situations.* For example, it is wise to use probes that relate to a learning goal in the teacher's curriculum and use his or her students' work to examine students' ideas and consider adjustments to instruction.

4. *Enable teachers to develop new ways of teaching that fit with their new understanding.* It is not enough for teachers to improve their own understanding of the content of a probe or to become aware of the misconceptions their students have. Teachers must consider how they will make the content accessible to their students and how they will change their teaching practices or lessons to help students develop conceptual understanding.

5. *Engage teachers in a continuous process of improvement.* Regular examination of student work from the probes (as well as the Teacher Notes that follow each probe) brings to light new problems related to teaching and learning, develops new understandings about content and pedagogy, and encourages teachers to make modifications to their lessons or try new instructional strategies.

Introduction

Nine Suggestions for Using Probes as Assessment for Teacher Learning

Volume 2 in this series addressed ways to use probes as assessment *for* learning to *promote* student learning and *inform* instruction. It also provided a variety of suggestions for embedding probes in instruction. Here, we offer suggestions for ways to use the probes in this book and in Volumes 1 and 2 as assessments *for teacher learning.*

1. **Always do the probe yourself.** Before giving the probe to your students, take it yourself and think about your prior experiences in school and in everyday situations that may be contributing to your response. Think about when and how you learned the content of the probe and whether you once had misconceptions or still have them. Note any difficulties you had responding to the probe. Being metacognitive about your own knowledge and experiences can help you understand what your students or other teachers might have thought or have experienced while taking the probe. It also points out areas of conceptual difficulty in which you may be helped by further professional development, university courses, or the support of a knowledgeable colleague or university partner.

2. **Use the Teacher Notes.** The Teacher Notes section that follows each probe contains a wealth of information that can deepen your understanding of the content of the probe and its curricular and instructional implications. Examine the Teacher Notes *before* looking at the student work to sharpen your analytical lens and to anticipate what you might find when you examine students' thinking. Use the Teacher Notes as references in your discussion of student work with other teachers in order to move beyond opinion and speak from a common knowledge base.

3. **Examine student work in collaborative structures.** Collaborative learning environments are good settings in which to examine the student work from the probes and to discuss the accompanying Teacher Notes. Structures that promote collegial learning include professional learning communities (PLCs), study groups, mentoring, instructional coaching, teacher research teams, and inquiry and reflective practice groups. The ongoing nature of these structures allows teachers to reconvene after making modifications or trying out new strategies and receive feedback from their colleagues on the formative decisions they made.

4. **Embed the probes into existing professional development.** Using the probes for teacher learning does not have to be a stand-alone professional learning experience. The probes can be used to examine student thinking within a

variety of content-focused workshops and institutes as well as within embedded professional development strategies, such as lesson study (Lewis 2002), teacher research (Roberts, Bove, and Van Zee 2007), curriculum topic study (Keeley 2005), study groups (Murphy and Lick 2001), and curriculum implementation, content immersion, and demonstration lessons (Loucks-Horsley et al. 2003).

5. **Select specific areas to focus on.** Looking at student work from the probes can initially be daunting. The responses to the probes provide such a treasure trove of data that it is sometimes hard to know what to focus on. Table 1 shows some of the things you might look for.

6. **Examine student thinking across grade spans.** An aspect of these probes that distinguishes them from summative assessments is that they are designed to be used across multiple grade levels. Sometimes, a single probe can be used at all levels—elementary, middle, and high school. For example, the "Is It a Solid?" probe shown in Figure 2 (p. 8) may elicit misconceptions about solids that may develop as early as first grade in a unit on solids and liquids and continue through high school if not addressed. Sometimes, the ideas do not change much from one grade span to the next; students just get a lot more "scientific" and their ideas become more muddled with other concepts they are learning as they move up through the grades.

Table 1. What to Look For in Students' Responses to Probes

Areas for Analysis	What to Look For
Concepts and ideas	Number of students choosing a selected response (use tallies); groups of students using similar explanations
Use of terminology	Confusion of everyday words with their scientific meaning; appropriate use of scientific terminology
Transfer of learning	Ability to apply ideas across contexts or in new situations
Prior knowledge or experience	Ideas that students bring to their learning; experiences students may have had that impact their ideas
Sophistication level	Grade levels at which the students' ideas are typically developed
Reasoning	Types of rules or justifications students use to support their ideas
Ability to explain	Students' ability to write or verbalize an explanation

Figure 2. "Is It a Solid?" Probe

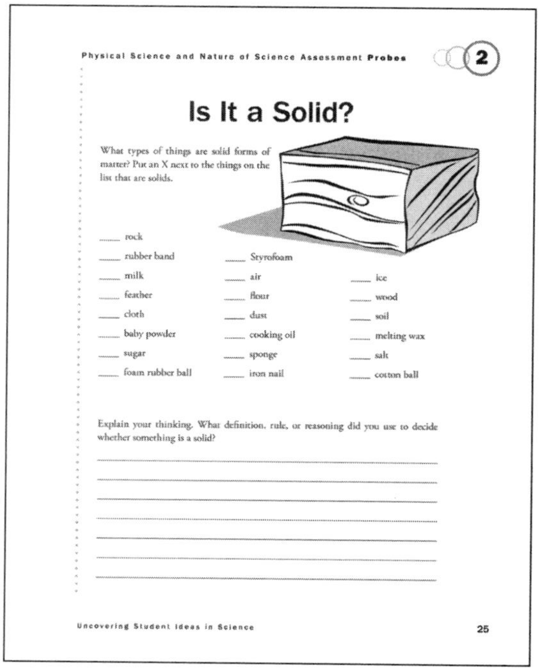
Physical Science and Nature of Science Assessment Probes 2

Is It a Solid?

What types of things are solid forms of matter? Put an X next to the things on the list that are solids.

___ rock
___ rubber band
___ milk
___ feather
___ cloth
___ baby powder
___ sugar
___ foam rubber ball
___ Styrofoam
___ air
___ flour
___ dust
___ cooking oil
___ sponge
___ iron nail
___ ice
___ wood
___ soil
___ melting wax
___ salt
___ cotton ball

Explain your thinking. What definition, rule, or reasoning did you use to decide whether something is a solid?

Uncovering Student Ideas in Science 25

There is tremendous value in analyzing students' responses to the same probe with colleagues from other grade levels. Because learning goals become more complex as students progress in school, it is important to take into account the targeted concept and learning goal for the particular grade you teach, including its level of sophistication. Here are some questions to ask yourself before and after you examine student work at various grade levels and read the Teacher Notes:

- How do my district's curricular learning goals at my grade level match the national learning goals described in the Teacher Notes?
- Which parts of my district's learning goals does the probe address?
- How well do my students know the learning goals related to a probe that come before my grade level? Are there learning goals that come after my grade level that they have had opportunities to achieve?
- To what extent does the Curricular and Instructional Considerations section described in the Teacher Notes match my classroom or district context? To what extent are my students familiar with the context of the probe, the correct response, and the distracters (wrong choices)? Should I tell students to ignore some of the distracters?
- Do any of the age-related research findings match the ages of students in my classes?
- When I look at the student work, do I find that any of my students' ideas are among the research findings in the Teacher Notes?
- What seems to be most problematic for students? Is this indicative of one grade span or is it seen across grades? What patterns do I notice?
- If students hold similar ideas, how do their explanations differ or remain the same across grade spans?
- Taking into account the types of ideas my students have and their reasoning, which points in the Suggestions for Instruction and

Assessment section in the Teacher Notes might be appropriate for me to use?

- What other curricular or instructional actions might I need to take?

7. **Categorize types of ideas.** Researchers and science educators have categorized students' science ideas in various ways. As you examine your students' ideas, try matching them to the following categories of ideas adapted from *Science Teaching Reconsidered* (NRC 1997):

   - **Scientific ideas.** These are the accurate conceptions that a scientifically literate person would have. Scientifically accurate ideas can range from basic, precursor understandings to sophisticated, complex ideas, depending on the developmental level of the student.

   - **Preconceptions.** These are popular conceptions based on everyday experiences. Often, they take root even before students have been taught scientific ideas. For example, students think the phases of the Moon are caused by the shadow of the Earth on the Moon. This conception is often rooted in students' everyday experience of seeing how part of an object is shaded when something blocks the light shining on the object.

   - **Conceptual misunderstandings.** These misunderstandings arise when students are taught scientific concepts in a way that does not challenge them to confront their beliefs or in a way that fails to connect disparate pieces of knowledge. For example, even though students have been taught the idea of the water cycle, they may believe that evaporated water moves immediately upward to a cloud. In their minds, they see the water going up to the clouds rather than existing in the air around them. They may correctly use terms like *evaporation* and *condensation* and know that water can exist as water vapor, yet fail to understand that water vapor is in the air around us. Another example is recanting an incorrect definition of matter but not accepting the idea that air is matter. Misrepresentations also lead to conceptual misunderstandings, such as the exaggerated elliptical orbit of the Earth around the Sun that leads to misconceptions about seasons. Overgeneralizations and undergeneralizations—for example, all animals have fur and legs and only shiny or smooth objects reflect light—are also categorized as conceptual misunderstandings.

   - **Nonscientific beliefs.** These are views learned by students from nonscientific sources, such as religious or mythical teachings, and

Introduction

pseudosciences, such as astrology. For example, in religious instruction, some students learn through literal interpretation of the Bible about the short time in which the Earth and the organisms that inhabit it were created. Scientific evidence, however, shows that the Earth was created about 5 billion years ago and life began with simple, one-celled organisms. The result has been considerable controversy in teaching certain aspects of science, such as the origins of the Earth or life and the Earth's and life's evolution.

- **Vernacular misconceptions.** These arise from the use of scientific words and phrases that mean one thing in everyday life and another in a scientific context. An example is the word *theory,* which means a "hunch" to some people but to scientists means a well-established, thoroughly tested idea. Sometimes the way words are used implies something other than what was intended; for example, "heat rises" and "the Sun moves across the sky" may imply that heat is the physical entity that rises rather than the air or that the Sun is actually the object that is moving around the Earth. Sometimes a scientific word is misused, as in the phrase "a hard candy melts in your mouth." Melt-

ing is a physical process. The candy does not melt, it dissolves. As a result of the misuse of the word *melt* students have a hard time distinguishing between melting and dissolving. Likewise, misuse of the term *zero gravity*—rather than the correct term, *microgravity*—has led students to think there are instances where there is no gravity acting on an object in space.

- **Factual misconceptions.** These are inaccuracies that may be taught and learned in school or home and are retained unchallenged throughout adulthood. For example, the notions that lightning never strikes twice in the same place, that water in a bathtub drain swirls in the opposite direction in the Southern Hemisphere than it does in the Northern Hemisphere, or that the blood in your veins is blue are all incorrect. Such ideas may remain part of many students' and adults' beliefs because they have been (incorrectly) taught them.

8. **Crunch the data and create classroom profiles and graphs.** Classroom profiles provide a written record of students' misconceptions for the teacher to analyze, share with other teachers, and refer to in monitoring conceptual change over time (Shapiro 1994). The classroom profile breaks each distracter down into the

Figure 3. Classroom Profile for "Thermometer" Probe (Grade 7)

Response Choices	Students' Supporting Explanations for Their Choices	# of Students
1. (Jean-Paul) "The hot water pushed it up."	• "The hot water creates a force that pushes the liquid up." • Repeats the answer choice with no further explanation.	2 1 **Total:** 3
2. (Pita) "The mass of the red liquid increased."	• "When the liquid gets hot from the water, it gets heavier and expands."	1 **Total:** 1
3. (Jonathan) "The heat inside the thermometer rises."	• "The red liquid is in the form of heat and it rises." • Explains how heat rises when it's warm but doesn't mention the red liquid. • Repeats the answer choice with no further explanation.	2 3 1 **Total:** 6
4. (Jimena) "The air inside the thermometer pulls it up."	• "Air pressure makes it go up." • "The air gets warm and creates a vacuum, which sucks the liquid up." • "The warm air creates a force that pulls up on the liquid."	1 1 1 **Total:** 3
5. (Molly) "The molecules of the red liquid are further apart."	• "As the liquid warms up, the space between the molecules increases." • "The liquid is turning to a gas so the molecules are further apart."	1 1 **Total:** 2
6. (Greta) "The number of molecules in the red liquid increased."	• "The molecules break apart when the liquid gets hot, making more molecules." • "There is more liquid so there are more molecules."	1 1 **Total:** 2
7. (Keanu) "The molecules of the red liquid are getting bigger."	• "The heat makes the molecules expand [also includes 'grow' and 'stretch.']" • "Things expand or get bigger when they warm up [but doesn't mention molecules]." • Gave no explanation—just "that's how thermometers work." • Gave no further explanation—left blank.	4 2 1 1 **Total:** 8

Introduction

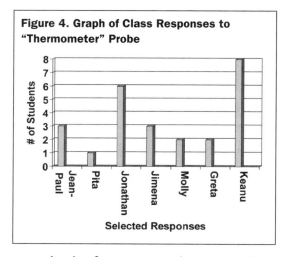

Figure 4. Graph of Class Responses to "Thermometer" Probe

kinds of reasoning students use and the number of students who used similar reasoning for that response. The exercise in grouping similar responses is an excellent professional development exercise for teachers that models similar ways that researchers code responses. Teachers can also include an additional column with the names of students who shared each idea. Classroom profiles can be shared with students (with names removed); most students show great interest in knowing what other students think and how the other students' thinking compares to their own. Creating graphs and charts to visually show the tallies of students' responses is another way of sharing data with students; the visual tallies are also useful in teacher collaborative groups. Figure 3 (p. 11) shows a classroom profile for a seventh-grade class's response to the probe "Thermometer" (on p. 33 of this book). Figure 4 shows a graph of the student responses from the probe.

9. **Read and discuss professional literature.** Further your understanding of the content related to the probe, students' ideas, and effective strategies and activities to use to teach for conceptual change by selecting and reading professional literature from the related NSTA publications list provided for each probe. In addition, you may want to read the references cited, including the full research papers. Also, search *www.curriculumtopicstudy.org.* This database regularly posts new research articles on students' misconceptions and other resources for teacher professional development related to the topic of each probe.

Whatever probe or probes you decide to use in your classroom, remember the outcomes will be threefold. First, you will learn a lot about your students; this new information will lead to modifications in how and what you teach. Second, you will learn a lot about standards and research-based teaching and learning that applies to all students. And third, you will learn the value in sharing the probes, your student data, your inquiries into practice, and your new learning with other teachers.

References

Atkin, M., and J. Coffey. 2003. *Everyday assessment.* Arlington, VA: NSTA Press.

Carlson, M., G. Humphrey, and K. Reinhardt. 2003. *Weaving science inquiry and continuous assessment.* Thousand Oaks, CA: Corwin Press.

Keeley, P. 2005. *Science curriculum topic study: Bridging the gap between standards and practice.* Thousand Oaks, CA: Corwin Press.

Keeley, P. 2008. *Science formative assessment: 75 practical strategies for linking assessment, instruction, and learning.* Thousand Oaks, CA: Corwin Press.

Keeley, P., F. Eberle, and L. Farrin. 2005. *Uncovering student ideas in science, vol. 1: 25 formative assessment probes.* Arlington, VA: NSTA Press.

Keeley, P., F. Eberle, and J. Tugel. 2007. *Uncovering student ideas in science, vol. 2: 25 more formative assessment probes.* Arlington, VA: NSTA Press.

Lewis, C. 2002. *Lesson study: A handbook of teacher-led instructional change.* Philadelphia: Research for Better Schools.

Loucks-Horsley, S., N. Love, K. Stiles, S. Mundry, and P. W. Hewson. 2003. *Designing professional development for teachers of science and mathematics.* Thousand Oaks, CA: Corwin Press.

Murphy, C., and D. Lick. 2001. *Whole-faculty study group: Creating student-based professional development.* Thousand Oaks, CA: Corwin Press.

National Research Council (NRC). 1997. *Science teaching reconsidered.* Washington, DC: National Academy Press.

Roberts, D., C. Bove, and E. Van Zee, eds. 2007. *Teacher research: Stories of learning and growing.* Arlington, VA: NSTA Press.

Shapiro, B. 1994. *What children bring to light: A constructivist perspective on children's learning in science.* New York: Teachers College Press.

Thompson, C., and J. Zueli. 1999. The frame and the tapestry: Standards-based reform and professional development. In *Teaching as the learning profession: Handbook of policy and practice,* eds. L. Darling-Hammond and G. Sykes, 341–375. San Francisco: Jossey Bass.

Physical Science and Nature of Science Assessment Probes

Physical Science and Nature of Science Assessment Probes
Concept Matrix

Related Science Concepts	Pennies	Is It a Solid?	Thermometer	Floating Balloon	Hot and Cold Balloons	Mirror on the Wall	Batteries, Bulbs, and Wires	Apple on a Desk	Rolling Marbles	Dropping Balls	Is It a Theory?	Doing Science	What Is a Hypothesis?
				Physical Science								Nature of Science	
Acceleration										✓			
Atom	✓												
Balanced Forces								✓					
Circular Motion									✓				
Complete Circuit							✓						
Conservation of Matter					✓								
Density				✓									
Electric Circuit							✓						
Electricity							✓						
Experiment												✓	
Force								✓	✓	✓			
Gas				✓	✓								
Gravity								✓		✓			
Hypothesis											✓		✓
Inertia									✓				
Kinetic Molecular Theory			✓	✓	✓								
Liquid		✓											
Mass				✓	✓					✓			
Mirrors						✓							
Nature of Science											✓	✓	✓
Newton's First Law									✓	✓			
Properties of Matter	✓	✓		✓	✓								
Reflection						✓							
Scientific Inquiry												✓	✓
Scientific Law											✓		
Scientific Method												✓	✓
Solid		✓											
Theory											✓		
Thermal Expansion			✓										
Thermometer			✓										
Weight				✓	✓					✓			

Pennies

A shiny new penny is made up of atoms. Put an X next to all the things on the list that describe the atoms that make up the shiny new penny.

___ hard   ___ soft

___ solid   ___ copper-colored

___ very small   ___ has mass

___ always moving   ___ do not move   ___ cold

___ warm   ___ shiny   ___ dull

___ made of smaller particles

___ contains mostly empty space

Describe your thinking about the atoms that make up the penny. Explain why you selected the things on the list as ways to describe atoms.

_____

_____

_____

_____

_____

_____

_____

_____

_____

_____

_____

Pennies

Teacher Notes

Purpose

The purpose of this assessment probe is to elicit students' ideas about the properties of atoms. The probe is designed to determine whether students can distinguish between the microscopic properties of an atom and the macroscopic properties of a substance or object made up of atoms.

Related Concepts

atom, properties of matter

Explanation

Five items on the list make up the best response: very small, has mass, always moving, made of smaller particles, and contains mostly empty space. Atoms are the smallest particles of matter that make up the substances zinc and copper (elements) in the penny. Most pen-

nies circulating today are made up of 97.5% zinc and 2.5% copper. If a penny were made of pure copper, it would contain about 2.4×10^{24} copper atoms. This indicates that the size of an individual atom is very small. A penny weighs only about 2 g. With over 10^{24} atoms in a penny, this shows that the mass of an individual atom is extremely small, yet it still has mass. Atoms and molecules are always in motion and reach a minimum of motion in very extreme, cold conditions. Since the penny is a solid, the atoms that make up the penny are in a fixed position and can only move by vibrating. An atom is mostly empty space. Even though the zinc and copper atoms make up the solid substance of the penny, the atoms themselves are mostly empty space. They consist of a small, dense nucleus surrounded by electrons that move in an area of space about a trillion

times larger in volume than the nucleus, making the total atom mostly empty space. The atom is made up of even smaller particles—protons and neutrons in the nucleus and electrons found outside the nucleus. Scientists have discovered even smaller particles that make up the protons and neutrons. Properties such as hard, solid, copper-colored, shiny, cold (or warm) describe the macroscopic properties of the substances (zinc and copper) or object (penny) and are not the properties of the individual atoms.

Curricular and Instructional Considerations

Elementary Students

At the elementary school level, students describe the properties of materials, objects, and familiar substances, like water. The focus is on observable and measurable properties of macroscopic matter.

Middle School Students

At the middle school level, students transition from focusing on the properties of objects and materials to the properties of substances. They develop an understanding of the atom as the smallest unit of matter that has mass and takes up space. They begin to distinguish between states of matter by using the idea of position and movement of atoms and molecules. However, the idea of empty space within an atom and between atoms is difficult for students at this age because they still tend to view matter as a continuous substance.

High School Students

At the high school level, students learn about the physical and chemical properties of atoms. They learn to distinguish between the macroscopic properties of elements and the microscopic properties of the atoms that make up elements. At this level, they learn about subatomic particles and the architecture of an atom. Their deepening understanding of kinetic molecular theory, introduced in middle school, helps them recognize that atoms and molecules are constantly moving and reach a minimum of motion in temperatures approaching absolute zero (0° Kelvin).

Administering the Probe

Because this probe targets ideas related to the properties of atoms, it is most suitable for middle school and high school grades. Consider showing students a shiny new penny and, although they do not need to know the composition of new pennies for this probe, you can explain that pennies today are not made entirely of copper.

Related Ideas in *National Science Education Standards* (NRC 1996)

K–4 Properties of Objects and Materials

- Objects have many observable properties, including size, weight, shape, color, temperature, and the ability to react with other substances. Those properties can be measured using tools, such as rulers, balances, and thermometers.

9–12 Structure of Atoms

★ Matter is made of minute particles called *atoms,* and atoms are composed of even smaller components. These components have measurable properties, such as mass and electric charge. Each atom has a positively charged nucleus surrounded by negatively charged electrons.

9–12 Structure and Properties of Matter

• Atoms interact with one another by transferring or sharing electrons that are farthest from the nucleus. These outer electrons govern the chemical properties of the element.

• Solids, liquids, and gases differ in the distances and angles between molecules or atoms and therefore the energy that binds them together. In solids, the structure is nearly rigid; in liquids, molecules or atoms move around each other but do not move apart; and in gases, molecules or atoms move almost independently of each other and are mostly far apart.

Related Ideas in *Benchmarks for Science Literacy* (AAAS 1993)

K–2 Structure of Matter

• Objects can be described in terms of the materials they are made of (clay, cloth, paper, etc.) and their physical properties (color, size, shape, weight, texture, flexibility, etc.).

3–5 Structure of Matter

• Materials may be composed of parts that are too small to be seen without magnification.

6–8 Structure of Matter

• All matter is made up of atoms, which are far too small to see directly through a microscope. The atoms of any element are alike but are different from atoms of other elements. Atoms may stick together in well-defined molecules or may be packed together in large arrays. Different arrangements of atoms into groups compose all substances.

★ Atoms and molecules are perpetually in motion. In solids, the atoms are closely locked in position and can only vibrate.

9–12 Structure of Matter

★ Atoms are made of a positive nucleus surrounded by negative electrons.

★ The nucleus, a tiny fraction of the volume of an atom, is composed of protons and neutrons, each almost 2,000 times heavier than an electron. The number of positive protons in the nucleus determines what an atom's electron configuration can be and so defines the element.

• Scientists continue to investigate atoms and have discovered even smaller constituents of which neutrons and protons are made.

Related Research

• Middle school and high school students are deeply committed to a theory of continuous matter. Although some students may think

---

★ Indicates a strong match between the ideas elicited by the probe and a national standard's learning goal.

that substances can be divided up into small particles, they do not recognize the particles as building blocks, but as formed of basically continuous substances under certain conditions (AAAS 1993, p. 336).

- Students of all ages show a wide range of beliefs about the nature and behavior of particles, including a lack of appreciation of the very small size of particles (AAAS 1993).

- Some students, when recognizing the minute size of atoms, reason that because atoms are so small they have zero or negligible mass (Driver et al. 1994).

- Although some students can depict the orderly arrangement of atoms or molecules in a solid, they have difficulty recognizing the vibration of the particles (Driver et al. 1994).

- Several studies of students' initial conception of an atom show that they perceive it either as "a small piece of material" or the "ultimate bit of material obtained when a portion of material is progressively subdivided." Such "bits" are thought to vary in size and shape and possess properties similar to the properties of the parent material. For example, some students consider atoms of a solid to have all or most of the macro properties that they associate with the solid, such as hardness, hotness/coldness, color, and state of matter (Driver et al. 1994, p. 74).

- Children's naive view of particulate matter is based on a "seeing is believing" principle in which they tend to use sensory reasoning. Being able to accommodate a scientific particle model involves overcoming cognitive difficulties of both a conceptual and perceptive nature (Kind 2004).

Suggestions for Instruction and Assessment

- Be explicit in developing the idea that any property of a material is a result of the arrangement of the particles, not a result of the individual particles having that property.

- Ask students to draw what they think an atom and a group of atoms look like. Their representations will vary and can be used as starting points for discussions about the properties of things we cannot see.

- Do not assume students will recognize the difference between properties of atoms and properties of substances. After teaching about properties of atoms and molecules, provide an opportunity for students to use a graphic organizer to compare and contrast microscopic and macroscopic properties at a substance and atomic/molecular level. Repeat this using an example of a solid, liquid, and gaseous substance.

- Be up-front with students about the difficulty in conceptualizing small particles like atoms and molecules. Explain how scientists have been trying to understand atoms for the last 2,000 years, and it was not until the early 19th century that the idea of atoms was accepted. It took more than a century after that to understand the structure of atoms, which is still being studied by scientists today. If it took scientists this long to understand the particulate nature of matter, then do not expect students to change their models overnight (Kind 2004).

- Have students imagine they are wearing "atomic spectacles" that allow them to "see" atoms. Show them a shiny penny, nickel, or dime and ask them what the atoms look like. Encourage them to draw the atoms. Have them discuss the differences and similarities between their ideas and drawings, further probing for ideas about color or continuous matter.

- Use analogies to depict the very small size of atoms in relation to the total volume of an atom.

- Demonstrate how something that is non-continuous can look continuous depending on the magnitude of our view. For example, show students a sponge and have them examine the many holes. Stand far enough away so that students cannot see the individual holes. The sponge will look like a continuous block of matter, even though it is full of holes. Relate this to looking at the penny without powers of magnification that would allow one to see at the atomic level.

- Representations of atoms often lead to students' misconceptions. Use the PRISMS (Phenomena and Representations for Instruction of Science in Middle School) website at *http://prisms.mmsa.org* to find examples of atomic representations, along with their likely instructional effectiveness, that can be used in teaching about atoms.

Related NSTA Science Store Publications and NSTA Journal Articles

American Association for the Advancement of Science (AAAS). 1993. *Benchmarks for science literacy.* New York: Oxford University Press.

American Association for the Advancement of Science (AAAS). 2001. *Atlas of science literacy.* Vol. 1, "atoms and molecules map," 54–55. Washington, DC: AAAS.

Driver, R., A. Squires, P. Rushworth, and V. Wood-Robinson. 1994. *Making sense of secondary science: Research into children's ideas.* London and New York: RoutledgeFalmer.

Hazen, R., and J. Trefil. 1991. *Science matters: Achieving scientific literacy.* New York: Anchor Books.

Keeley, P. 2005. *Science curriculum topic study: Bridging the gap between standards practice.* Thousand Oaks, CA: Corwin Press.

National Research Council (NRC). 1996. *National science education standards.* Washington, DC: National Academy Press.

Related Curriculum Topic Study Guide

(Keeley 2005)

"Particulate Nature of Matter (Atoms and Molecules)"

References

American Association for the Advancement of Science (AAAS). 1993. *Benchmarks for science literacy.* New York: Oxford University Press.

Driver, R., A. Squires, P. Rushworth, and V. Wood-Robinson. 1994. *Making sense of secondary sci-

ence: Research into children's ideas. London and New York: RoutledgeFalmer

Keeley, P. 2005. *Science curriculum topic study: Bridging the gap between standards and practice.* Thousand Oaks, CA: Corwin Press.

Kind, V. 2004. *Beyond appearances: Students' misconceptions about basic chemical ideas.* 2nd ed.

Durham, England: Durham University School of Education. Also available online at *www.chemsoc.org/pdf/LearnNet/rsc/miscon.pdf.*

National Research Council (NRC). 1996. *National science education standards.* Washington, DC: National Academy Press.

Is It a Solid?

What types of things are solid forms of matter? Put an X next to the things on the list that are solids.

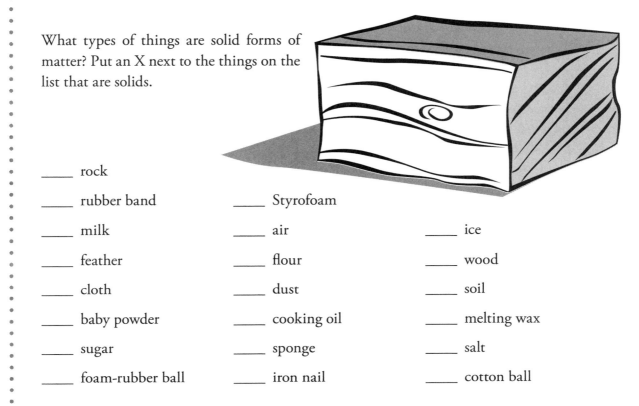

_____ rock

_____ rubber band   _____ Styrofoam

_____ milk   _____ air   _____ ice

_____ feather   _____ flour   _____ wood

_____ cloth   _____ dust   _____ soil

_____ baby powder   _____ cooking oil   _____ melting wax

_____ sugar   _____ sponge   _____ salt

_____ foam-rubber ball   _____ iron nail   _____ cotton ball

Explain your thinking. What definition, rule, or reasoning did you use to decide whether something is a solid?

_____

_____

_____

_____

_____

_____

_____

Is It a Solid?

. . . .

Teacher Notes

Purpose

The purpose of this assessment probe is to elicit students' ideas about solids. The probe is designed to reveal the macroscopic and/or microscopic properties students use to decide whether a material is a solid.

Related Concepts

liquid, properties of matter, solid

Explanation

The best responses are rock, rubber band, Styrofoam, ice, feather, flour, wood, cloth, dust, soil, baby powder, sugar, sponge, salt, foam rubber ball, iron nail, and cotton ball. The items on the list that are liquids are milk, cooking oil, and melting wax. The only gas on the list is air. A

solid is a material in which the atoms or molecules are in a fixed position and can only vibrate in place. The atoms or molecules in a liquid are more loosely connected; they are able to slide past one another but are not independent of each other as in a gas. However, there are some solids such as graphite (a form of carbon) that have the useful property of having layers of carbon slide over each other. Sometimes the atoms or molecules of a liquid gain enough energy to form a gas and move independently of each other. Gases have a random atomic or molecular organization and much more space between the atoms or molecules than those of a solid or liquid, which enables gases to be compressed.

From a macroscopic level, solids generally maintain their shape and have a definite

volume. Individual particles do not slide over each other, as in a liquid, which accounts for why liquids can assume the shape of their container and can be poured. Some collections of tiny parts of solid materials, such as matter in granular or powder form (sand or flour), assume the shape of their container and can be poured. However, this is because each granule or speck of powder is an individual, tiny piece of solid and not because the atoms or molecules that make up the substance are sliding over each other. It is the collection of these tiny pieces that behaves in this way, much like filling or pouring a jar full of solid marbles.

The word *solid* is often used in an everyday sense to imply something hard or not "airy." The rubber band, Styrofoam, foam-rubber ball, sponge, and cotton ball are soft or airy but they still fit the definition of a solid as a material in which the molecules are in a fixed position and vibrate in place, regardless of how hard, soft, compact, or airy the object is. However, foams may be considered neither solid nor liquid if one recognizes they are mixtures where a gas is finely dispersed within a solid.

Curricular and Instructional Considerations

Elementary Students

At the elementary level, students describe the properties of materials or objects and classify them as solids, liquids, or gases. Their definition of a solid is based on macroscopic properties such as an object keeping its shape and having a definite volume. The students' mac-

roscopic definition of a liquid is based on the object taking the shape of its container and having a definite volume.

Middle School Students

At the middle school level, students transition from focusing on the macroscopic properties of solids, liquids, and gases to explaining states of matter in terms of the position and arrangement of the atoms or molecules.

High School Students

At the high school level, students deepen their understanding of the behavior of solids, liquids, and gases based on the objects' position, arrangement, and motion. They also explore the characteristics of the fourth state of matter, plasma. They examine non-Newtonian fluids and other unusual materials, such as putties, pastes, dough, foams, and gels, including colloidal mixtures in which solid particles are mixed with water, and explain their behavior at a particle level to determine whether they are considered solids or liquids or mixtures of two different states of matter.

Administering the Probe

Eliminate objects from the list that students are not familiar with. Consider providing a visual prop, either a picture or an actual object, for each item on the list. For example, if students do not know what a foam-rubber ball is, you might show them a familiar Nerf ball toy. This probe can also be administered as a card-sort activity (Keeley 2008).

Related Ideas in *National Science Education Standards* (NRC 1996)

K–4 Properties of Objects and Materials

★ Materials can exist in different states, as a solid, liquid, or gas.

9–12 Structure and Properties of Matter

★ Solids, liquids, and gases differ in the distances and angles between molecules or atoms and therefore the energy that binds them together. In solids, the structure is nearly rigid; in liquids, molecules or atoms move around each other but do not move apart; and in gases, molecules or atoms move almost independently of each other and are mostly far apart.

Related Ideas in *Benchmarks for Science Literacy* (AAAS 1993)

K–2 Structure of Matter

• Objects can be described in terms of the materials they are made of (clay, cloth, paper, etc.) and their physical properties (color, size, shape, weight, texture, flexibility, etc.).

3–5 Structure of Matter

• Heating and cooling causes changes in the properties of materials.

6–8 Structure of Matter

★ Atoms and molecules are perpetually in motion. Increased temperature means greater average energy of motion, so most substances expand when heated. In solids, the atoms are closely locked in position and can only vibrate. In liquids, the atoms or molecules have higher energy, are more loosely connected, and can slide past one another; some molecules may get enough energy to escape into a gas. In gases, the atoms or molecules have still more energy and are free of one another, except during occasional collisions.

9–12 Structure of Matter

• An enormous variety of biological, chemical, and physical phenomena can be explained by changes in the arrangement and motion of atoms and molecules.

Related Research

• Students of all ages show a wide range of beliefs about the nature and behavior of particles, including a difficulty in appreciating the intrinsic motion of particles in solids, liquids, and gases (AAAS 1993).

• A study of children's ideas about solids conducted with Israeli children ages 5–13 showed that younger children tend to associate solids with rigid materials (Stavy and Stachel 1984). They regard powders as liquids, and any nonrigid materials, such as a sponge or a cloth, as being somewhere in between a solid and liquid (Driver et al. 1994).

★ Indicates a strong match between the ideas elicited by the probe and a national standard's learning goal.

- Students' explanation of powders as liquids is often "because they can be poured." Reasons for nonrigid objects as being neither solid nor liquid is because they "are soft," "crumble," or "can be torn." Thus children characterize the state of matter of a material according to its macroscopic appearance and behavior with the result that solids are associated with hardness, strength, and an inability to bend (Driver et al. 1994).

- By age 11, students tend to regard a powder as being an intermediate state, rather than a liquid (Driver et al. 1994).

- Stavy and Stachel (1984) concluded that children can classify liquids more easily than they can solids, perhaps because liquids are less varied in their physical characteristics (Kind 2004).

- Although some students can depict the orderly arrangement of atoms or molecules in a solid, they have difficulty recognizing the vibration of the particles (Driver et al. 1994).

- Students tend to recognize materials like metals and wood as being solids. However, students have difficulty categorizing materials that are not hard or rigid as solids. Fifty percent of 12- to 13-year-olds classified nonrigid solids like dough, sponge, sand, and sugar differently from coins, glass, or chalk. They suggest that "the easier it is to change the shape or state of the solid, the less likely it is to be included in the group of solids" (Kind 2004, p. 6).

- Children's naive view of particulate matter is based on a "seeing is believing" principle in which they tend to use sensory reasoning. Being able to accommodate a scientific particle model involves overcoming cognitive difficulties of both a conceptual and perceptual nature (Kind 2004).

Suggestions for Instruction and Assessment

- Be aware that you may have to probe deeper than just asking students to describe solids. For elementary school students, solids are just the step into the door of states of matter. Focusing instruction on just solids is not getting into the more conceptual knowledge of the states of matter.

- When investigating solids, elementary school students should be exposed to a wide variety of solids, including rigid and soft materials, porous and nonporous materials, and solids made up of small particles, such as sand and sugar.

- With elementary school students, it is important not to overemphasize ease of flow as a property of liquids, since solids such as sand and salt seem to have this property. Instead, concentrate on the individual particles with a hand lens and pencil point, observing that each grain has its own form and keeps that form when pushed or squashed, unlike a drop of water. Have students observe how the solid particles form a heap when poured out onto a flat surface, which liquids do not. By beginning with coarse particles such as sand or salt, children can go on to see that finer powders like flour and baby powder are also solids (Wenham 2005).

- Students should also observe and describe

the behavior of collections of larger pieces, such as marbles, sugar cubes, or wooden blocks (which can, for example, be poured out of a container), and consider that the collections may have new properties that the individual pieces do not (AAAS 1993, p. 76). Relate this to smaller pieces in materials like powders, sand, salt, and sugar.

- Researchers suggest that upper elementary school students (around age 11) have an opportunity to develop the idea that a powder is composed of small pieces of a solid. However, researchers warn that, when subsequently learning the particulate theory of solids, students may wrongly infer that the theoretical particles are "powder grains." Therefore, it is suggested that, before they learn particulate theory, students should be capable of classifying materials according to a scientific view of the states of matter (Driver et al. 1994, p. 79).

- It is important to develop the concept of particulate solids before investigating colloids and suspensions.

Related NSTA Science Store Publications and NSTA Journal Articles

Adams, B. 2006. Science shorts: All that matters. *Science and Children* (Sept.): 53–55.

American Association for the Advancement of Science (AAAS). 1993. *Benchmarks for science literacy.* New York: Oxford University Press.

Buchanan, K. 2005. Idea bank: Oobleck and beyond. *The Science Teacher* (Dec.): 52–54.

Driver, R., A. Squires, P. Rushworth, and V. Wood-

Robinson. 1994. *Making sense of secondary science: Research into children's ideas.* London and New York: RoutledgeFalmer.

Keeley, P. 2005. *Science curriculum topic study: Bridging the gap between standards and practice.* Thousand Oaks, CA: Corwin Press.

National Research Council (NRC). 1996. *National science education standards.* Washington, DC: National Academy Press.

Ontario Science Center. 1995. *Solids, liquids, and gases: Starting with science series.* Toronto: Kids Can Press.

Related Curriculum Topic Study Guides

(Keeley 2005)

"Particulate Nature of Matter" (Atoms and Molecules)
"Solids"
"States of Matter"

References

American Association for the Advancement of Science (AAAS). 1993. *Benchmarks for science literacy.* New York: Oxford University Press.

Driver, R., A. Squires, P. Rushworth, and V. Wood-Robinson. 1994. *Making sense of secondary science: Research into children's ideas.* London and New York: RoutledgeFalmer

Keeley, P. 2005. *Science curriculum topic study: Bridging the gap between standards and practice.* Thousand Oaks, CA: Corwin Press.

Keeley, P. 2008. *Science formative assessment: 75 practical strategies for linking assessment, instruction, and learning.* Thousand Oaks, CA: Corwin Press.

Kind, V. 2004. *Beyond appearances: Students' misconceptions about basic chemical ideas.* 2nd ed. Durham, England: Durham University School of Education. Also available online at *www. chemsoc.org/pdf/LearnNet/rsc/miscon.pdf.*

National Research Council (NRC). 1996. *National science education standards.* Washington, DC: National Academy Press.

Stavy, R., and Stachel, D. 1984. *Children's ideas about "solid" and "liquid."* Tel Aviv, Israel: Israeli Science Teaching Center, School of Education, Tel Aviv University.

Wenham, M. 2005. *Understanding primary science: Ideas, concepts, and explanations.* 2nd ed. Thousand Oaks, CA: Sage Publications.

Thermometer

Mr. Martinez placed a thermometer in a jar of very hot water. His students watched what happened to the thermometer. Immediately the level of the red liquid in the thermometer went up. His students disagreed about why the red liquid in the thermometer rose when the thermometer was placed in hot water. This is what they said:

Jean-Paul: "The hot water pushed it up."

Pita: "The mass of the red liquid increased."

Jonathan: "The heat inside the thermometer rises."

Jimena: "The air inside the thermometer pulls it up."

Molly: "The molecules of the red liquid are further apart."

Greta: "The number of molecules in the red liquid increased."

Keanu: "The molecules of the red liquid are getting bigger."

Which student do you most agree with? _____

Explain why you think that student has the best explanation.

_____

_____

_____

_____

_____

_____

Thermometer

Teacher Notes

Purpose

The purpose of this assessment probe is to elicit students' ideas about thermal expansion. The probe is designed to find out whether students attribute expansion of the space between molecules to the rise of the liquid in a thermometer.

Related Concepts

kinetic molecular theory, thermal expansion, thermometer

Explanation

Molly has the best answer: A thermometer is a closed system. It operates on the principle that the fluid inside it (usually alcohol or mercury) expands when heated and contracts when cooled. When the bulb is in contact with a warm object such as the hot water, energy from the hot water is transferred to the liquid inside the bulb. The molecules of the red liquid, in this case alcohol with a red dye added, gain energy and increase their motion as the faster-moving molecules bump up against and push the slower-moving molecules. This causes the molecules to move farther apart, and as a result, the alcohol inside the thermometer occupies more space as it expands. In order to occupy more space, the alcohol has to rise in the narrow tube. It is this increased motion and collisions of the molecules inside the very narrow tube that accounts for the rise of the alcohol.

Curricular and Instructional Considerations

Elementary Students

At the elementary school level, students use thermometers to measure the temperature of objects and materials. At this level they are developing the procedural skills of using a thermometer. They are not expected to know how a thermometer works.

Middle School Students

At the middle school level, students continue to use thermometers. They learn how a thermometer works and should be able to explain how it operates at a substance level—most substances expand or contract when they are heated or cooled. Some students can begin to use particle ideas to explain why a substance expands when heated and contracts when cooled and connect that to what happens inside a thermometer. At this stage they also recognize water as an anomaly to the idea that substances expand when heated and contract when cooled, noting that when water cools to form ice, it expands.

High School Students

At the high school level, students deepen their understanding of kinetic molecular theory and relate the thermometer phenomenon to particle ideas about thermal expansion. At this level, they are expected to be able to explain how a thermometer works based on the expansion or contraction of the liquid due to increasing or decreasing space between the molecules as a result of increased or decreased motion when energy is gained or lost by the molecules.

Administering the Probe

This probe can be demonstrated for students using a red alcohol thermometer or performed in small groups with appropriate safety precautions. The word *volume* is intentionally not used to describe the "liquid going up" in order to probe for younger students' ideas related to the visible increase in the height of the liquid without having their lack of understanding of what volume is interfering with their ideas about the phenomenon. For middle school and high school students who understand the concept of volume, you can replace "His students disagreed about why the red liquid in the thermometer rose when the thermometer was placed in hot water" with "…why the volume of red liquid in the thermometer increased when the thermometer was placed in hot water."

Related Ideas in *National Science Education Standards* (NRC 1996)

K–4 Abilities Necessary to Do Scientific Inquiry

- Employ simple equipment and tools (thermometers) to gather data and extend the senses.

K–4 Properties of Objects and Materials

- Objects have many observable properties, including size, weight, shape, color, temperature, and the ability to react with other substances. Those properties can be measured using tools, such as rulers, balances, and thermometers.

5–8 Abilities Necessary to Do Scientific Inquiry

- Use appropriate tools (thermometers) and techniques to gather, analyze, and interpret data.

5–8 Transfer of Energy

- Energy is a property of many substances and is associated with heat. Energy is transferred in many ways.
- Heat moves in predictable ways, flowing from warmer objects to cooler ones, until both objects reach the same temperature.

9–12 Conservation of Energy and the Increase in Disorder

- ★ Heat consists of random motion and the vibrations of atoms, molecules, and ions. The higher the temperature, the greater the atomic or molecular motion.

Related Ideas in *Benchmarks for Science Literacy* (AAAS 1993)

3–5 Structure of Matter

- Heating and cooling cause changes in the properties of materials. Many kinds of changes occur faster under hotter conditions.

6–8 Structure of Matter

- ★ Atoms and molecules are perpetually in motion. Increased temperature means greater average energy of motion, so most substances expand when heated.

6–8 Energy Transformations

- Heat can be transferred through materials by the collisions of atoms or across space by radiation.
- Energy appears in different forms. Heat energy is in the disorderly motion of molecules.

9–12 Transformations of Energy

- Heat energy in a material consists of the disordered motions of its atoms or molecules.

Related Research

- Some students tend to regard liquids as continuous (nonparticulate) and static (Driver et al. 1994).
- In an Australian study of 25 children ages 8–11, children were asked how they thought a thermometer worked (Appleton 1985). About one-third of the children suggested the thermometer "was sensitive to heat," or that it "was made to go to the right number." Other suggestions involved pressure, pushing, or heat rising (Driver et al. 1994).

Suggestions for Instruction and Assessment

- Have students research how to make a thermometer and then have them build one. Students should demonstrate the use of their thermometers and explain how they work at a substance level and a particle level (if they are ready to use atomic/ molecular reasoning).
- Use the rising level of liquid in a thermometer as a plausible phenomenon to develop

---

★ Indicates a strong match between the ideas elicited by the probe and a national standard's learning goal.

the idea that most substances expand when heated.

- Trace the transfer of thermal energy in a thermometer from the hot water to the glass to the alcohol. Have students draw a visual representation of the transfer of energy between molecules.

- Have students draw pictures to show what happens to the liquid in a thermometer at the particle level when the bulb comes in contact with hot material. Use the drawings (whiteboards work well for this) to discuss students' ideas about conduction, the particle nature of matter, and kinetic molecular theory.

- Use the analogy of playing pool to illustrate what happens when molecules collide and transfer energy. When a pool cue ball hits a rack of pool balls, it transfers energy and the balls it hits spread out.

- Help the students who chose Jonathan's response to understand how some words in science are used incorrectly. For example, the common phrase "heat rises" is incorrect. It is the warm air or water that rises, not the heat.

- Probe students' reasoning further for each of the distracters chosen and challenge their ideas. For example, Greta's idea can be challenged with conservation of matter reasoning, including the idea of a closed system in which no additional molecules can get into the thermometer.

- Relate expansion of the liquid in a thermometer to expansion of a metallic object. A metal ball and ring apparatus, available

through most science supply stores, demonstrates how a metal expands when heated by showing how the ball passes through the ring before the ball is heated, but not after it is heated. Have students connect this phenomenon to what happens inside the thermometer.

Related NSTA Science Store Publications and NSTA Journal Articles

American Association for the Advancement of Science (AAAS). 1993. *Benchmarks for science literacy.* New York: Oxford University Press.

American Association for the Advancement of Science (AAAS). 2007. *Atlas of science literacy.* Vol. 2, "states of matter map," 58–59. Washington, DC: AAAS.

Driver, R., A. Squires, P. Rushworth, and V. Wood-Robinson. 1994. *Making sense of secondary science: Research into children's ideas.* London and New York: RoutledgeFalmer.

Keeley, P. 2005. *Science curriculum topic study: Bridging the gap between standards and practice.* Thousand Oaks, CA: Corwin Press.

National Research Council (NRC). 1996. *National science education standards.* Washington, DC: National Academy Press.

Robertson, W. 2002. *Energy, Stop Faking It! Finally Understanding Science So You Can Teach It.* Arlington, VA: NSTA Press.

Related Curriculum Topic Study Guides

(Keeley 2005)

"Heat and Temperature"

"Particulate Nature of Matter" (Atoms and Molecules)

References

American Association for the Advancement of Science (AAAS). 1993. *Benchmarks for science literacy.* New York: Oxford University Press.

Appleton, K. 1985. Children's ideas about temperature. *Research in Science Education* 15: 122–126.

Driver, R., A. Squires, P. Rushworth, and V. Wood-Robinson. 1994. *Making sense of secondary science: Research into children's ideas.* London and New York: RoutledgeFalmer

Keeley, P. 2005. *Science curriculum topic study: Bridging the gap between standards and practice.* Thousand Oaks, CA: Corwin Press.

National Research Council (NRC). 1996. *National science education standards.* Washington, DC: National Academy Press.

Floating Balloon

Shemal has an uninflated balloon. He fills the uninflated balloon with a gas and ties it closed. When he lets go, the balloon floats up into the sky. Shemal wonders what happens to the mass of the uninflated balloon compared to the inflated, floating balloon. What do you think? Circle the answer that best matches your thinking.

The floating balloon has more mass.

The floating balloon has less mass.

The mass of the uninflated balloon and the floating balloon is the same.

Describe your thinking. Provide an explanation for your answer.

_____

_____

_____

_____

_____

_____

_____

_____

_____

_____

Floating Balloon

Teacher Notes

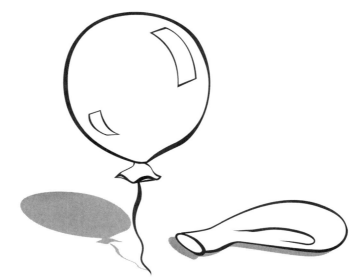

Purpose

The purpose of this assessment probe is to elicit students' ideas about the mass of a gas. The probe is designed to reveal whether students recognize that an uninflated balloon will increase in mass when inflated with a gas, even though the balloon intuitively seems lighter when it floats in the air.

Related Concepts

density, gas, kinetic molecular theory, mass, properties of matter, weight

Explanation

The best response is A: The floating balloon has a mass greater than the mass of the uninflated balloon. Gases are a form of matter and thus have mass (or weight) and take up space. Putting a gas (such as helium in this case) in a balloon adds mass, even though some students intuitively think the balloon may now be lighter because it floats. Balloons filled with helium rise because they are less dense (the total mass-to-volume ratio) than the surrounding air and not because they have a lesser mass than an uninflated balloon. Students tend to confuse density with mass. In this case, the mass of the gas-filled balloon has increased while its density has decreased with the increase in volume. A balloon filled with helium has a lesser mass (and weighs less) than a balloon blown up to the same size with air because it is less dense. However, this problem is not about comparing density, but rather comparing mass. The

balloon filled with helium has a greater mass than the uninflated balloon because additional mass (matter in the form of a gas) was added to the "empty" balloon.

Curricular and Instructional Considerations

Elementary Students

At the elementary school level, students describe the properties of materials or objects and classify them as solids, liquids, or gases. Their experiences with solids and liquids are based on matter they can see. Gases are more difficult for them to understand as they have not yet developed a particulate notion of matter. However, before students proceed to middle school, it is important for elementary school students to understand that gases are matter and that they have weight (and mass).

Middle School Students

At the middle school level, students transition from focusing on the macroscopic properties of solids, liquids, and gases to explaining states of matter in terms of the position, arrangement, and motions of the atoms or molecules. Compared with students in elementary school grades, middle school students have more experience investigating gases. At this level, they should understand the idea that gases are made of molecules that have mass (and weight). As they investigate density, they discover that some gasses, like helium, are less dense than the surrounding air. However, they need to be able to distinguish between density and mass

in order to reason why an object's mass (not a substance's mass) can increase while its density decreases. At this level, students begin to use the number of atoms and molecules to explain the conservation of matter or change in mass. If the number of atoms or molecules in an object remains the same before and after a change, then the mass stays the same. Conversely, if additional molecules or atoms are added to an object, such as putting gas into an "empty" balloon, then the mass increases.

High School Students

At the high school level, students deepen their understanding of gases by learning about the gas laws and behavior of fluids. However, they tend to hold on to their earlier ideas about the mass (and weight) of a gas if they are not confronted with their preconceptions. They may confuse density-related ideas with the comparison of masses (empty vs. gas-filled balloon) by not accounting for the change in volume in this phenomenon.

Administering the Probe

Students should be familiar with the rising of a helium-filled balloon. If not, consider bringing in a helium-filled balloon and an empty balloon of the same size and shape and ask students to think about the mass of each. If using this probe with younger elementary students, consider using the words *empty* instead of *uninflated* and *weight* instead of *mass* so that students' unfamiliarity with the concept of mass does not interfere with their ability to answer this question and explain their reasoning.

Related Ideas in *National Science Education Standards* (NRC 1996)

K–4 Properties of Objects and Materials

- Objects have many observable properties, including size, weight, shape, color, temperature, and the ability to react with other substances. Those properties can be measured using tools, such as rulers, balances, and thermometers.
- Materials can exist in different states, as a solid, liquid, or gas.

9–12 Structure and Properties of Matter

- Matter is made of minute particles called *atoms*, and atoms are composed of even smaller components. These components have measurable properties, such as mass.

Related Ideas in *Benchmarks for Science Literacy* (AAAS 1993)

K–2 Structure of Matter

- Objects can be described in terms of the materials they are made of (clay, cloth, paper, etc.) and their physical properties (color, size, shape, weight, texture, flexibility, etc.).

3–5 Structure of Matter

- Materials may be composed of parts that are too small to be seen without magnification.

3–5 The Earth

- Air is a substance that surrounds us, takes up space, and whose movement we feel as wind.

6–8 Structure of Matter

- Equal volumes of different substances usually have different weights.
- The idea of atoms explains the conservation of matter: If the number of atoms stays the same no matter how they are rearranged, then their total mass stays the same. (Conversely, if the number of atoms changes, then the object's mass changes.)

9–12 Structure of Matter

- An enormous variety of biological, chemical, and physical phenomena can be explained by changes in the arrangement and motion of atoms and molecules.

Related Research

- Students may believe that matter does not include gases or that gases are weightless materials (AAAS 1993).
- Researchers have suggested that one of the reasons students fail to recognize gases as having weight or mass is because one of their most common experiences with gases is with those that tend to rise or float (such as the helium). This view is supported by studies that show that children ages 9–13 tend to predict that gases have negative weight, such that when a gas like helium is added to a balloon, it will weigh less. Students believe that the

more gas that is added to a container, the lighter the container becomes (Driver et al. 1994).

- Students at the end of elementary school and beginning of middle school may be at different points in their conceptualization of a "theory" of matter. Although some third graders may start seeing weight as a fundamental property of all matter, many students in sixth and seventh grade still appear to think of weight simply as "felt weight"—something whose weight they cannot feel is considered to have no weight at all (AAAS 1993, p. 336).

- Many researchers have noted that students do not initially seem to be aware that air and other gases are a type of "material" and thus have properties, such as weight or mass, like other materials (Driver et al. 1994).

- The idea that air or gas has mass is not obvious to children. Yet, when it is taught, it is a concept children can acquire easily and remember (Sere 1985).

- Gases pose special difficulties for children since the ones they commonly experience, like air and helium, are invisible. It is suggested that this invisibility prevents students from developing a scientific conception of a gas. Explicit instruction is needed for children to understand the properties of a gas, including properties like mass and weight. This is in contrast to solids and liquids where students tend to learn about them intuitively (Kind 2004).

Suggestions for Instruction and Assessment

- Before students can describe the properties of a gas, they must first accept a gas as being a substance or matter.

- Provide opportunities for students to use fundamental physical dimensions such as quantity, mass, volume, pressure, and temperature to describe the state of a gas (Sere 1985).

- Provide students with opportunities to enclose a gas (such as blowing up a balloon with air) and find the mass. Compare the mass of the empty balloon with the mass of the balloon filled with air. This can also be demonstrated by balancing two empty balloons on the end of a balanced meterstick or other lever. Blow up one balloon and reattach it to show the gain in mass.

- Use particle drawings to help students reason what happens when the uninflated balloon is filled with air. Have students draw particles of air in the "empty" balloon and then draw particles of air put into the balloon and tied off so the air does not escape. Have them compare the number of particles and then ask if an increase in the number of air particles increases the mass. Once students agree that the mass increases because air particles were added and each particle adds mass, then ask what would happen if other gases were used. Establishing the idea that all gases add particles, which add mass, may lead to helping students understand why a balloon's total mass does not decrease when helium is added.

- For older students who have developed a concept of density, challenge them to show how the mass of an object containing a gas can increase while its density decreases. Have them connect this idea to the phenomenon described in the probe.

Related NSTA Science Store Publications and NSTA Journal Articles

Adams, B. 2006. Science shorts: All that matters. *Science and Children* (Sept.): 53–55.

American Association for the Advancement of Science (AAAS). 1993. *Benchmarks for science literacy.* New York: Oxford University Press.

Driver, R., A. Squires, P. Rushworth, and V. Wood-Robinson. 1994. *Making sense of secondary science: Research into children's ideas.* London and New York: RoutledgeFalmer.

Keeley, P. 2005. *Science curriculum topic study: Bridging the gap between standards and practice.* Thousand Oaks, CA: Corwin Press.

National Research Council (NRC). 1996. *National science education standards.* Washington, DC: National Academy Press.

Ontario Science Center. 1995. *Solids, liquids, and gases: Starting with science series.* Toronto: Kids Can Press.

Robertson, W. 2005. *Air, water, and weather: Stop Faking It! Finally Understanding Science So You Can Teach It.* Arlington, VA: NSTA Press.

Sadler, T., T. Eckart, J. Lewis, and K. Whitley. 2005. Tried and true: It's a gas! An exploration of the physical nature of gases. *Science Scope* (Nov./Dec.): 12–14.

Related Curriculum Topic Study Guide
(Keeley 2005)
"Behavior and Characteristics of Gases"

References

American Association for the Advancement of Science (AAAS). 1993. *Benchmarks for science literacy.* New York: Oxford University Press.

Driver, R., A. Squires, P. Rushworth, and V. Wood-Robinson. 1994. *Making sense of secondary science: Research into children's ideas.* London and New York: RoutledgeFalmer.

Keeley, P. 2005. *Science curriculum topic study: Bridging the gap between standards and practice.* Thousand Oaks, CA: Corwin Press.

Kind, V. 2004. *Beyond appearances: Students' misconceptions about basic chemical ideas.* 2nd ed. Durham, England: Durham University School of Education. Also available online at *www.chemsoc.org/pdf/LearnNet/rsc/miscon.pdf.*

National Research Council (NRC). 1996. *National science education standards.* Washington, DC: National Academy Press.

Sere, M. 1985. The gaseous state. In *Children's ideas in science,* eds. R. Driver, E. Guesne, and A. Tiberghien, 105–123. Milton Keynes, UK: Open University Press.

Hot and Cold Balloons

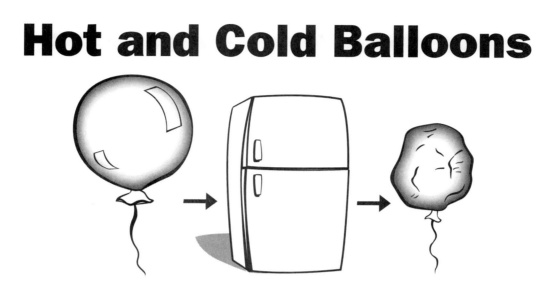

Moira filled a balloon with air. She tightly tied the balloon so no air could get in or out of the balloon. She kept the balloon in a warm room. An hour later she put the balloon in a cold freezer. When she took the balloon out 30 minutes later, it was still tied tightly shut. No air escaped from the balloon; however, the balloon had shrunk.

Moira wondered if the mass of the balloon (including the air inside it) also changed. Circle the answer that best matches your thinking.

A The mass of the warm balloon is less than the mass of the cold balloon.

B The mass of the warm balloon is greater than the mass of the cold balloon.

C The mass of the warm balloon is the same as the mass of the cold balloon.

Describe your thinking. Provide an explanation for your answer.

_____

_____

_____

_____

_____

_____

_____

Hot and Cold Balloons

Teacher Notes

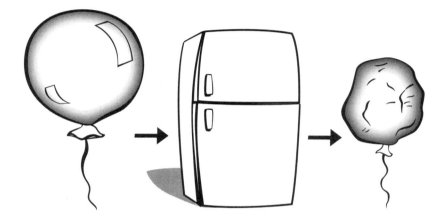

Purpose

The purpose of this assessment probe is to elicit students' ideas about conservation of matter. The probe is designed to reveal whether students recognize that the mass of a warm gas in a closed system is the same after it has been cooled, even though the volume it occupies has decreased.

Related Concepts

conservation of matter, gas, kinetic molecular theory, mass, properties of matter, weight

Explanation

The best response is C: The mass of the warm balloon is the same as the mass of the cold balloon. In the warm balloon, the gas molecules are free of each other and moving rapidly. They collide with each other and with the wall of the balloon exerting pressure, which gives the balloon its size (volume). As the balloon cools in the freezer, the air molecules transfer energy to the surrounding freezer. The molecules of air are still free of each other but they do not move as rapidly or as far apart and their collisions with the balloon wall are not as forceful. As a result, the volume of the balloon containing the cool air decreases. The conservation of matter principle explains why the masses are the same. Mass is the measure of the amount of matter in an object, material, or substance. Heat and cold only speed up or slow down the motion of molecules. Changing the temperature of the air inside a closed system does not change the mass because the number of molecules remains the same. The balloon is sealed so nothing can get in or out.

Curricular and Instructional Considerations

Elementary Students

At the elementary school level, students describe the properties of materials or objects and classify them as solids, liquids, or gases. Their experiences with solids and liquids are based on matter they can see. Gases are more difficult for them to understand, because they have not yet developed a particulate notion of matter. However, before proceeding to middle school, it is important for students at the elementary school level to understand that gases are matter and they have weight. At this age level, conservation of matter is taught using phenomena with pieces that are observable, such as parts of objects. The shrinking balloon phenomena used in this probe is appropriate at the observational level for elementary school students, but asking students to explain what happens in terms of the particles should wait until they are ready to use a particulate model.

Middle School Students

At the middle school level, students transition from focusing on the macroscopic properties of solids, liquids, and gases to explaining states of matter in terms of the position, arrangement, and motions of the atoms or molecules. Compared with elementary school grades, middle school students have more experiences investigating gases. At this level, they should understand the idea that gases are made of molecules that have mass (and weight). They develop the idea of a closed system and can use that idea

to reason conservation of matter–related phenomena, although understanding conservation of matter in a gas context is more difficult.

High School Students

At the high school level, students deepen their understanding of gases by learning about the gas laws. They use Charles's law to explain what happens to the volume of a gas when the temperature changes. At this grade level, students are expected to be able to use the conservation of matter principle to explain a variety of changes within a closed system. However, they tend to hold on to their earlier ideas about the mass (and weight) of a gas if not confronted with their preconceptions.

Administering the Probe

Consider demonstrating this phenomenon with a balloon. If a freezer is not available, put the balloon outside in the cold, in a refrigerator, or in an ice chest. Make sure students understand that for the purpose of this probe, the balloon is sealed, although in reality some air can escape.

Related Ideas in *National Science Education Standards* (NRC 1996)

K–4 Properties of Objects and Materials

- Objects have many observable properties, including size, weight, shape, color, temperature, and the ability to react with other substances. Those properties can be measured using tools, such as rulers, balances, and thermometers.

- Materials can exist in different states, as a solid, liquid, or gas.

9–12 Structure and Properties of Matter

- Solids, liquids, and gases differ in the distances and angles between molecules or atoms and therefore the energy that binds them together. In solids, the structure is nearly rigid; in liquids, molecules or atoms move around each other but do not move apart; and in gases, molecules or atoms move almost independently of each other and are mostly far apart.

9–12 Conservation of Energy and the Increase in Disorder

- Heat consists of random motion and the vibrations of atoms, molecules, and ions. The higher the temperature, the greater the atomic or molecular motion.

Related Ideas in *Benchmarks for Science Literacy* (AAAS 1993)

K–2 Structure of Matter

- Objects can be described in terms of the materials they are made of (clay, cloth, paper, etc.) and their physical properties (color, size, shape, weight, texture, flexibility, etc.).

3–5 Structure of Matter

- Heating and cooling cause changes in the

properties of materials.

- No matter how parts of an object are assembled, the weight of the whole object is always the same as the sum of the parts.

3–5 The Earth

- Air is a substance that surrounds us, takes up space, and whose movement we feel as wind.

6–8 Structure of Matter

- Atoms and molecules are perpetually in motion. Increased temperature means greater average energy of motion, so most substances expand when heated. In solids, the atoms are closely locked in position and can only vibrate. In liquids, the atoms or molecules have higher energy, are more loosely connected, and can slide past one another; some molecules may get enough energy to escape into a gas. In gases, the atoms or molecules have still more energy and are free of one another, except during occasional collisions.

★ No matter how substances within a closed system interact with one another or how they combine or break apart, the total mass of the system remains the same. The idea of atoms explains the conservation of matter: If the number of atoms stays the same, no matter how they are rearranged, then their total mass stays the same.

9–12 Structure of Matter

- An enormous variety of biological, chemical, and physical phenomena can be ex-

---

★ Indicates a strong match between the ideas elicited by the probe and a national standard's learning goal.

plained by changes in the arrangement and motion of atoms and molecules.

Related Research

- Students may believe that matter does not include gases or that gases are weightless materials (AAAS 1993).
- Many researchers have noted that students do not initially seem to be aware that air and other gases are a type of "material" and thus have properties, such as weight or mass, like other materials (Driver et al. 1994).
- Research shows that some students have a difficult time conserving matter in a closed container when a gas is involved and the volume of the container changes. They confuse volume with quantity (Sere 1985). This can also be explained by the intuitive rule, "more A, more B" noted by Stavy and Tirosch (1995). Since the volume of the room temperature balloon is larger, students reason that the balloon has more matter, thus more mass.
- The idea that air or gas has mass is not obvious to children. Yet, when it is taught, it is a concept children can acquire easily and remember (Sere 1985).
- Some students believe a warmed gas weighs less than the same gas that is cooler (Driver et al. 1994).

Suggestions for Instruction and Assessment

- Have students carry out investigations to test their ideas. Use their findings to engage them in resolving the discrepancy be-

tween their prediction and ideas and their findings. However, be careful in humid climates that additional mass from condensation of water vapor in the air is not added to the balloon when it comes out of the freezer.

- Explicitly teach the concept of a closed and open system. Link conservation of matter to changes that happen in a closed system.
- Gases pose special difficulties for children because the gases they commonly experience, like air and helium, are invisible. It is suggested that this invisibility prevents students from developing a scientific conception of a gas. Explicit instruction is needed for children to understand the properties of a gas, including properties like mass and weight. This is in contrast to solids and liquids where students tend to learn about them intuitively (Kind 2004).
- Provide other opportunities to compare changes in volume of a gas with temperature, such as slipping a balloon over a flask and then heating it, observing the balloon as it expands. Ask students to explain what happens to the mass of the total system before and after heating.
- Encourage students to draw a "particle picture" of what is happening inside the balloon in both situations. Use their drawings to probe deeper into their understanding of the numbers and motion of the particles.
- Combine this probe with probes in Volume 1 (Keeley, Eberle, and Farrin 2005) to further examine students' ideas about conservation of matter during a physical change.

Related NSTA Science Store Publications and NSTA Journal Articles

American Association for the Advancement of Science (AAAS). 1993. *Benchmarks for science literacy.* New York: Oxford University Press.

Driver, R., A. Squires, P. Rushworth, and V. Wood-Robinson. 1994. *Making sense of secondary science: Research into children's ideas.* London and New York: RoutledgeFalmer.

Keeley, P. 2005. *Science curriculum topic study: Bridging the gap between standards and practice.* Thousand Oaks, CA: Corwin Press.

National Research Council (NRC). 1996. *National science education standards.* Washington, DC: National Academy Press.

Robertson, W. 2005. *Air, water, and weather: Stop Faking It! Finally Understanding Science So You Can Teach It.* Arlington, VA: NSTA Press.

Sadler, T., T. Eckart, J. Lewis, and K. Whitley. 2005. Tried and true: It's a gas! An exploration of the physical nature of gases. *Science Scope* (Nov./Dec.): 12–14.

Related Curriculum Topic Study Guides

(Keeley 2005)

"Behavior and Characteristics of Gases"

"Conservation of Matter"

References

American Association for the Advancement of Science (AAAS). 1993. *Benchmarks for science literacy.* New York: Oxford University Press.

Driver, R., A. Squires, P. Rushworth, and V. Wood-Robinson. 1994. *Making sense of secondary science: Research into children's ideas.* London and New York: RoutledgeFalmer.

Keeley, P. 2005. *Science curriculum topic study: Bridging the gap between standards and practice.* Thousand Oaks, CA: Corwin Press.

Keeley, P., F. Eberle, and L. Farrin. 2005. *Uncovering student ideas in science: 25 formative assessment probes.* Vol. 1. Arlington, VA: NSTA Press.

Kind, V. 2004. *Beyond appearances: Students' misconceptions about basic chemical ideas.* 2nd ed. Durham, England: Durham University School of Education. Also available online at *www.chemsoc.org/pdf/LearnNet/rsc/miscon.pdf.*

National Research Council (NRC). 1996. *National science education standards.* Washington, DC: National Academy Press.

Sere, M. 1985. The gaseous state. In *Children's ideas in science,* eds. R. Driver, E. Guesne, and A. Tiberghien, 105–123. Milton Keynes, UK: Open University Press.

Stavy, R., and D. Tirosch. 1995. *How students (mis) understand science and mathematics: Intuitive rules.* New York: Teachers College Press.

Mirror on the Wall

Adrienne placed a small, flat mirror flat against a wall. Standing close to the mirror, Adrienne could see her face from her eyebrows to her chin. Adrienne backed up five steps away from the mirror. Adrienne is now farther away from the mirror. How much of her face will Adrienne see in the mirror this time?

A She will see more of her face.

B She will see less of her face.

C She will see the same amount of her face.

Explain your thinking. Describe your reasoning or any experiences you have had with mirrors that helped you decide what Adrienne would see.

_____

_____

_____

_____

_____

_____

_____

_____

_____

_____

_____

Mirror on the Wall

Teacher Notes

Purpose

The purpose of this assessment probe is to elicit students' ideas about reflection of light. The probe can be used to examine how students use ideas about light to explain how we see objects in a mirror.

Related Concepts

mirrors, reflection

Explanation

The best response is C: She will still only see the same amount of her face, from her eyebrows to her chin. Despite the fact that we look into a mirror everyday and hundreds of times in a year, most people believe that the further away you are from a mirror, the more you will see of yourself. Our familiarity with mirrors clearly does not mean we understand how they work. This probe shows that experience is not always the best teacher. When light falls on a flat mirror it is reflected in a predictable way: the angle at which the light strikes the mirror (angle of incidence) and the angle at which it is reflected (angle of reflection) are the same. As long as the position of your eyes is in the same horizontal plane in relation to the mirror on the wall, it does not matter how far back or close you are to the mirror, you will still see the same image.

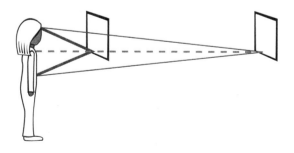

When you look into a mirror, you see the portions of your body from which light is reflected off the mirror and into your eyes. In the diagram on page 52, the person can see her head down to the middle of her chest. The light rays reflected to her eye from the bottom edge of the mirror determine the lowest visible portion of her body. The mirror reflects all light rays away at the same angle at which they arrived, so light rays from lower down on her body below the middle of her chest will be reflected to a point above her eyes, and will not be visible. When she steps back from the mirror, the light rays are reflected in the same manner as when she stood close to it. Although the light rays from it are reflected at a more shallow angle, the lowest visible portion of her body remains the same as when she was close to the mirror. The angle of light going in is equal to the angle of light reflected from the mirror, and thus the image stays the same regardless of how close or far back the mirror is.

Curricular and Instructional Considerations

Elementary Students

In the elementary school grades, students investigate reflection of light with mirrors. Their experiences are observational at this stage. Students are not expected to know how a mirror works. However, this probe is useful in helping students see that even though they may experience something every day and think they know what happens (i.e., believing they will see more of their face when they are further away from a mirror), their predictions do not always match their observations. Science can show us that what we think will happen, based on our everyday experience, is not always what actually happens, thus reinforcing the importance of testing our predictions rather than relying only on our experiences.

Middle School Students

At the middle school level, students learn how mirrors work and connect their expanding knowledge of light reflection to different types of objects that reflect light, such as mirrors. They can investigate how the angle at which light strikes and reflects off of a mirror into the eye determines what is seen in the mirror. However, even with instruction, students have difficulty understanding how an image appears in a mirror and is seen by the eye.

High School Students

At the high school level, students develop sophisticated ideas related to optics. They learn about how light interacts with different types of mirrors, such as concave and convex. However, as in the previous grades, their ideas about mirrors are strongly affected by their everyday experiences looking into mirrors.

Administering the Probe

You can model this scenario with a small rectangular mirror, about half the size of your face. Describe what is meant by placing the mirror at eye level and backing away from it.

Related Ideas in *National Science Education Standards* (NRC 1996)

K–4 Light, Heat, Electricity, and Magnetism

- Light travels in a straight line until it strikes an object. Light can be reflected by a mirror, refracted by a lens, or absorbed by the object.

5–8 Transfer of Energy

- Light interacts with matter by transmission (including refraction), absorption, or scattering (including reflection). To see an object, light from that object—either emitted by or scattered from it—must enter the eye.

Related Ideas in *Benchmarks for Science Literacy* (AAAS 1993)

3–5 Motion

- Light travels and tends to maintain its direction of motion until it interacts with an object or a material. Light can be absorbed, redirected, bounced back, or allowed to pass through. (Note: This is a new benchmark. It can be found in AAAS 2001, p. 65.)

6–8 Motion

- Something can be "seen" when light waves emitted or reflected by it enter the eye.
- Light acts like a wave in many ways. Waves can explain how light behaves. (Note: This

is a new benchmark. It can be found in AAAS 2001, p. 65.)

9–12 Motion

- Waves can superimpose on one another, bend around corners, reflect off surfaces, be absorbed by materials they enter, and change direction when entering a new material.

Related Research

- When children were asked if moving their position would change the position of an image on a mirror, over half thought that it would. Ninety percent of the children thought that moving back from a mirror would allow them to see more of themselves in the mirror (Driver et al. 1994).
- Difficulties in understanding how light travels contribute to students' misconceptions about how light interacts with mirrors and how light must enter the eye in order to see an object (Driver et al. 1994).
- Mirrors are typically used in curriculum units on light to demonstrate characteristics of light reflection. Several studies have shown that students have difficulty understanding how an image forms on a plane mirror (Shapiro 1994).
- The Annenberg/CPB Private Universe Project (1995) asked high school students a question similar to that asked in this probe. These students too thought that they would see more of themselves as they backed up from the mirror. The video shows the surprise they experienced when

they tried it and found they saw the same amount of their body.

Suggestions for Instruction and Assessment

- This probe can be tested by students by placing a small, flat rectangular mirror at eye level flat against a wall and having students back up from the mirror. Students are often quite surprised to find that their prediction that they would see more of themselves as they backed away from the mirror, which was based on their everyday experiences, did not match their result. Seize this opportunity to help students understand why testing a prediction is important in science, even when you think you know what the result will be. Experience is not always the best teacher! Sometimes actual results are unexpected. Be aware that some students are so convinced they will see more of themselves when they back up, that they sometimes actually think they do when they try it!

- With older students, pose the question, how big does a mirror need to be to see your whole body? After drawing reflection diagrams that explain why a mirror only needs to be half your height to see your full body, regardless of how far back or close you are standing, this problem can be linked to the probe scenario.

- Be aware that many students will still not accept the finding that the image in a plane mirror stays the same regardless of distance and may try to test it in the bathroom.

Some bathroom mirrors are placed on top of counters. The counter blocks light from the part of their body below the counter from being reflected to the mirror and then to the eye. When you back away from the counter, some of the light from the part of the body that was blocked by the counter is now able to reflect on to the mirror without obstruction and reflect back to your eye, thus allowing you to see more of your body than when you were standing close to the counter. Explain that the difference between this example and the example in the probe of the small, flat mirror on the wall is that there is no obstruction between the person and the small, flat mirror.

- To demonstrate the effect of the bathroom counter described above or other obstruction, place an 18- to 24-inch mirror on a table or counter and have students look into the mirror. Explain that the counter blocks light from the part of the body below the counter from being reflected to the mirror and then to the eye. Move the mirror to a wall. Have them observe how the light reflects without obstruction and reflects back to the eye, thus allowing the students to see more of their bodies than when they were standing close to the counter or table. Discuss the differences and why it seems as if backing up from a mirror in a bathroom makes it seem as if backing up from any mirror allows you to see more of yourself.

- Challenge students to find out why some mirrors, such as car mirrors, appear to reflect a much wider view.

- This probe can be used even if students are not learning about mirrors in the context of lessons about light reflection. The probe makes a case for why students should not rely solely on their everyday experiences when thinking about science ideas. Testing ideas is important because it sometimes reveals that what we strongly think, based on prior experiences, is not always the way the natural world works. Carefully testing our predictions helps us confront and develop new ideas related to understanding the natural world.

Related NSTA Science Store Publications and Journal Articles

American Association for the Advancement of Science (AAAS). 1993. *Benchmarks for science literacy.* New York: Oxford University Press.

American Association for the Advancement of Science (AAAS). 2001. *Atlas of science literacy.* Vol. 1, "waves," 64–65. Washington, DC: AAAS.

Driver, R., A. Squires, P. Rushworth, and V. Wood-Robinson. 1994. *Making sense of secondary science: Research into children's ideas.* London and New York: RoutledgeFalmer.

Keeley, P. 2005. *Science curriculum topic study: Bridging the gap between standards and practice.* Thousand Oaks, CA: Corwin Press.

Matkins, J., and J. McDonnough. 2004. Circus of light. *Science and Children* (Mar.): 50–54.

National Research Council (NRC). 1996. *National science education standards.* Washington, DC: National Academy Press.

Robertson, W. 2003. *Light, Stop Faking It! Finally Understanding Science So You Can Teach It.* Arlington, VA: NSTA Press.

Related Curriculum Topic Study Guide
(Keeley 2005)
"Visible Light, Color, and Vision"

References

American Association for the Advancement of Science (AAAS). 1993. *Benchmarks for science literacy.* New York: Oxford University Press.

American Association for the Advancement of Science (AAAS). 2001. *Atlas of science literacy.* Vol. 1, "waves," 64–65. Washington, DC: AAAS.

Driver, R., A. Squires, P. Rushworth, and V. Wood-Robinson. 1994. *Making sense of secondary science: Research into children's ideas.* London and New York: RoutledgeFalmer.

Keeley, P. 2005. *Science curriculum topic study: Bridging the gap between standards and practice.* Thousand Oaks, CA: Corwin Press.

National Research Council (NRC). 1996. *National science education standards.* Washington, DC: National Academy Press.

Private Universe Project. 1995. *The private universe teacher workshop series* [videotape]. South Burlington, VT: The Annenberg/CPB Math and Science Collection.

Shapiro, B. 1994. *What children bring to light: A constructivist perspective on children's learning in science.* New York: Teachers College Press.

Batteries, Bulbs, and Wires

Kirsten has a battery and a small bulb. She wonders how many strips of wire she will need to connect the battery and the bulb so that the bulb will light. What is the *smallest number* of wire strips Kirsten needs to make the bulb light up?

   A One strip of wire

   B Two strips of wire

   C Three strips of wire

   D Four strips of wire

Explain your thinking about how to light the bulb. Draw a picture to support your explanation.

Picture:

Batteries, Bulbs, and Wires

Teacher Notes

Purpose

The purpose of this assessment probe is to elicit students' ideas about complete circuits involving lightbulbs. The probe reveals whether students recognize the pathway of electricity in a complete circuit, including its path through a lightbulb, in order to light a bulb using only one strip of wire and a battery.

Related Concepts

complete circuit, electricity

Explanation

The best answer is A: One strip of wire. If you closely examine a flashlight bulb, you will see two small wires sticking up in the bulb that are connected by a very fine wire called a *filament*. The two wires on either side of the filament extend downward into the base of the bulb where

you cannot see them through the metal casing that surrounds the base of the bulb. One of these wires goes down to the very bottom of the base (the pointed end). The other wire is connected to the side of the metal base (sometimes the side is ridged so that it can screw into a socket). Knowing where these wires end up on the base of the bulb (the tip and the side) is necessary in order to use one wire to make a circuit that lights a bulb.

The battery, wire, and bulb need to be connected in such a way that it forms a complete circuit. To do this, hold the end of the wire against the negative terminal (the bottom of the battery or smooth end). The other end of the wire should touch or wrap around the side of the metal casing that forms the base of the lightbulb. With the wire wrapped around the metal side of the bulb and the other end

touching the bottom of the battery, touch the tip of the base of the lightbulb to the positive terminal (bumpy end) and the bulb will light.

The lightbulb lights with just one wire because the electricity flows out of the negative terminal (bottom of the battery), through the wire, into the wire that is attached to the side of the metal casing on the base of the bulb, up through that wire inside the bulb, across the filament, and down the other wire inside the bulb where it is attached to the point on the base of the bulb that touches the positive terminal (the bump) of the battery, completing a full circuit.

Be aware, however, that students can choose the best answer, "one wire," and still have an incorrect configuration of a circuit. For example, some students may touch one end of the wire to the battery and the other to the bulb, thinking the energy from the battery will flow through the wire to the bulb. Students who choose two wires as their answer may understand that a complete circuit is needed, but not understand the internal architecture of a lightbulb.

Curricular and Instructional Considerations

Elementary Students

"Batteries and bulbs" is a common instructional topic found in elementary school curriculum materials that help students acquire skills of inquiry while learning what a complete circuit is and about the architecture of a lightbulb. Building complete circuits is primarily observational at this level and provides students with an opportunity to systematically test out their ideas and make observations.

Middle School Students

In middle school, students build a variety of different circuits to trace the path of electricity. At this level, they are able to understand why the lightbulb is designed the way it is in order to build a circuit. They also begin to build an understanding of the direction in which the electric current flows. Students build on their earlier experiences with complete circuits to understand the transfer of energy that is involved in a circuit designed to accomplish a task such as lighting a lightbulb.

High School Students

Students at the high school level have begun the study of particles smaller than atoms, and their study may begin to link particles such as electrons to the notion of charge and to ideas of current and electric circuits. However, it is important to remember that many students may have had little experience with electric circuits themselves or that their experiences involved prefabricated circuit boards and light sockets that prevent them from seeing how the current flows through a lightbulb.

Administering the Probe

Make sure students know what you mean by a "strip" of wire. Show them a battery, a bulb, and a coil of wire. Cut a strip off that is long enough to make the circuit so they know what is meant by a strip of wire. Or, show a handful of cut strips, but do not show just one, because it might cue them to the answer. Note that if students choose more than one wire, it is not necessarily incorrect in terms of making a complete circuit. However the probe is not asking how many wires are needed to make a complete circuit but rather what the minimum number of wires is. It is important to emphasize drawing as a way of supporting the students' explanation in this probe.

Related Ideas in *National Science Education Standards* (NRC 1996)

K–4 Light, Heat, Electricity, and Magnetism

★ Electricity in circuits can produce light, heat, sound, and magnetic effects. Electric circuits require a complete loop through which an electric current can pass.

5–8 Transfer of Energy

• Electric circuits provide a means of transferring electric energy when heat, light, sound, and chemical changes are produced.

Related Ideas in *Benchmarks for Science Literacy* (AAAS 1993)

6–8 Structure of Matter

• Materials vary in how they respond to electric currents, magnetic forces, and visible light or other electromagnetic waves. (Note: This is a new benchmark. It can be found in AAAS 2007, p. 27.)

9–12 Forces of Nature

• Different kinds of materials respond differently to electric forces. In conducting materials such as metals, electric charges flow easily, whereas in insulating materials such as glass, they can move hardly at all.

Related Research

• Studies by Shipstone (1985), Arnold and Millar (1987), and Borges and Gilbert (1999) show that, before instruction, many K–8 students are not aware of the bipolarity of batteries and lightbulbs. They do not recognize the need for a complete circuit and have difficulty making a bulb light when provided with a battery and wires. Furthermore, even high school and university students have shown difficulty with this task (AAAS 2007).

• Many students will use a source-consumer model in which the battery gives something to the bulb. In this context, younger students will often draw a single wire going from the top of the battery (unipolar model) to the bulb. Another similar model

---

★ Indicates a strong match between the ideas elicited by the probe and a national standard's learning goal.

held by younger and older students involves two wires, each one going out of an end of the battery (bipolar model) and touching the bulb with the electricity going from the battery to the bulb in each wire (Driver et al. 1994).

- Some students will regard one wire as the "active" wire and the second wire as a "safety wire" (Driver et al. 1994).

Suggestions for Instruction and Assessment

- This probe lends itself to a follow-up inquiry investigation. Provide a battery, bulb, and single wire, and ask students to figure out whether it is possible to light a bulb with just one wire. Have them explain the path of the complete circuit once they figure out how to light the bulb.

- The metal casing around the base of a bulb prevents students from seeing what the structure of a lightbulb actually is. Before students can explain the complete circuit, they need to see the arrangement of the wires in the bulb. Use an enlarged cutaway diagram of a lightbulb to show students what the wires look like behind the casing. (Note: Do not break the glass of a lightbulb to show students the structure of a bulb. Students can be injured by shards of broken glass that may still be on the bulb casing.)

- Ask middle school and high school students to draw arrows to show the direction of the current. Sometimes their drawings with two wires will reveal a "clash model"

in which current flows from both ends of the battery and ends up lighting the bulb when the currents clash. This can provide further assessment information to inform instruction on how students think about electric current and complete circuits.

- Extend the probe to have students show how to light the bulb with two wires or with two batteries and one wire.

- Challenge students to find as many different ways as they can to light the bulb with one or two wires and explain the path of the current.

- Many students have trouble with batteries/bulbs/wires questions because they have only encountered them when they are part of a kit that includes "housing" for the batteries—with clips—so they are easier to manipulate and set up in a series. It is important to give students an opportunity to explore batteries, bulbs, and wires *without* the casings or clips so that they can understand how the current flows through each of the components.

Related NSTA Science Store Publications and NSTA Journal Articles

American Association for the Advancement of Science (AAAS). 1993. *Benchmarks for science literacy.* New York: Oxford University Press.

American Association for the Advancement of Science (AAAS). 2007. *Atlas of science literacy.* Vol. 2, "electricity and magnetism," 26–27. Washington, DC: AAAS.

Driver, R., A. Squires, P. Rushworth, and V. Wood-

Robinson. 1994. *Making sense of secondary science: Research into children's ideas.* London and New York: RoutledgeFalmer.

Keeley, P. 2005. *Science curriculum topic study: Bridging the gap between standards and practice.* Thousand Oaks, CA: Corwin Press.

National Research Council (NRC). 1996. *National science education standards.* Washington, DC: National Academy Press.

Robertson, W. 2005. *Electricity and magnetism: Stop Faking It! Finally Understanding Science So You Can Teach It.* Arlington, VA: NSTA Press.

Stein, M. 1998. Toying with science. *Science and Children* (Sept.): 35–39.

Related Curriculum Topic Study Guide

(Keeley 2005)

"Electrical Charge and Energy"

References

American Association for the Advancement of Science (AAAS). 1993. *Benchmarks for science literacy.* New York: Oxford University Press.

American Association for the Advancement of Science (AAAS). 2007. *Atlas of science literacy.* Vol. 2, "electricity and magnetism," 26–27. Washington, DC: AAAS.

Arnold, M., and R. Millar. 1987. Being constructive: An alternative approach to the teaching of introductory ideas in electricity. *International Journal of Science Education* 9 (5): 553–563.

Borges, A., and J. Gilbert. 1999. Mental models of electricity. *International Journal of Science Education* 21 (1): 95–117.

Driver, R., A. Squires, P. Rushworth, and V. Wood-Robinson. 1994. *Making sense of secondary science: Research into children's ideas.* London and New York: RoutledgeFalmer.

Keeley, P. 2005. *Science curriculum topic study: Bridging the gap between standards and practice.* Thousand Oaks, CA: Corwin Press.

National Research Council (NRC). 1996. *National science education standards.* Washington, DC: National Academy Press.

Shipstone, D. 1985. Electricity in simple circuits. In *Children's ideas in science,* eds. R. Driver, E. Guesne, and A. Tiberghien, 33–51. Milton Keynes, UK: Open University Press.

Apple on a Desk

Mrs. Canales pointed to an apple sitting on her desk. She asked her students to describe any forces acting on the apple. This is what some of her students said.

Archie: "The only force acting on the apple is air pressure."

Sam: "There is one force acting on the apple. Gravity is the force that pulls on the apple."

Soledad: "There are two forces: the desk pushes up on the apple and gravity pulls downward on the apple."

Misha: "There are many forces acting on the apple; but, it is the holding force in the apple that keeps it on the desk."

Tess: "There are no forces acting on the apple because the desk stops any forces from acting on it."

Which student do you most agree with? _____

Explain your thinking. What rule or reasoning did you use to decide if there were any forces acting on the apple?

_____

_____

_____

_____

_____

_____

Apple on a Desk

Teacher Notes

Purpose

The purpose of this assessment probe is to elicit students' ideas about forces. The probe is designed to find out whether students recognize that balanced forces act on a stationary object.

Related Concepts

balanced forces, force, gravity

Explanation

The best response is Soledad's: There are two forces: The desk pushes up on the apple and gravity pulls downward on the apple. Forces come in pairs. The force of gravity acting on the apple is the result of the matter in the Earth pulling on the matter in the apple. When the apple is set on the table, the table exerts a force upward on the apple equal to the force exerted downward, which is the pull on the apple. At

a microscopic level, when the apple is placed on the table, the individual molecules of the table's surface adjust their positions in much the same way that the individual springs in a bedspring mattress change position to support a sleeping person. This force on the apple exists because the apple is in contact with the surface of the table. When the apple is removed, the molecules of the table return to their original positions, as happens when a sleeping person rises from bed in the morning.

This force of the table on the apple is less obvious in our everyday experience. Forces such as that exerted by the table are difficult to detect or conceptualize because many of the structures we live among (tables, floors, walls, etc.) are rigid and thus show no apparent give and take when objects are placed on them. The fact that the apple is not moving indicates that

another force besides gravity must be present. In order for any object's motion to remain unchanged, all of the forces on that object must balance. In the case of the motionless apple, the downward gravitational force is balanced by an upward force supplied by the only other object in contact with the apple—the table. Air pressure also creates a force on the apple, but since air pushes on the apple almost equally in all directions, the effects of the air's force are not noticeable in this case.

Curricular and Instructional Considerations

Elementary Students

Elementary school instruction is primarily focused on describing the position and motion of objects and discovering the various kinds of forces that affect the motion of objects. Students develop the notion of forces as pushes and pulls on an object. Most of their learning about objects at rest is observational, with explanations coming later. They notice that things fall if not held up, and they later connect this to the words and concepts of *gravity* and *force,* including the notion that the Earth pulls on an object.

Middle School Students

Students at the middle school level engage in concrete experiences from which a more comprehensive understanding of force will develop in high school. Students develop the notion of balanced and unbalanced forces and describe the forces acting on objects. Instruction needs to include the idea that forces can be active or passive, because students at this age tend to equate force with motion and think there is no force acting on an object that is not moving.

High School Students

As students in high school begin to appreciate the particulate nature of matter and its minuteness of scale, they can also begin to understand the qualities of the electromagnetic forces that are dominant among atoms and their smaller particles. However, students may resist ideas about the importance and strength of these forces in their everyday experiences, especially as they relate to inanimate objects. This probe is useful in identifying students' ideas about forces, either before or after physics instruction.

Administering the Probe

This probe is most appropriate for middle and high school students. A visual prop such as an actual apple on your desk can be used to enhance the prompt. In addition to the explanation, consider asking students to draw a labeled diagram to show the forces and their directions.

Related Ideas in *National Science Education Standards* (NRC 1996)

K–4 Position and Motion of Objects

- The position and motion of objects can be changed by pushing or pulling.

5–8 Motion and Forces

★ If more than one force acts on an object along a straight line, then the forces will reinforce or cancel one another.

9–12 Motion and Forces

★ Whenever one object exerts force on another, a force equal in magnitude and opposite in direction is exerted on the first object.

★ Gravitation is a universal force that each mass exerts on any other mass.

Related Ideas in *Benchmarks for Science Literacy* (AAAS 1993)

K–2 Forces of Nature

• Things near the Earth fall to the ground unless something holds them up.

3–5 Forces of Nature

• The Earth's gravity pulls any object toward it without touching it.

6–8 Forces of Nature

• Every object exerts gravitational force on every other object.

9–12 Forces of Nature

• Gravitational force is an attraction between masses.

• Electromagnetic forces acting within and between atoms are vastly stronger than the gravitational forces acting between the atoms.

9–12 Motion

★ Whenever one thing exerts a force on another, an equal amount of force is exerted back on it.

Related Research

• Students tend to distinguish between active objects and objects that support or block or otherwise act passively, such as a table. Students tend to recognize the active actions as forces but often do not consider passive actions to be forces. Teaching students to integrate the concept of passive support into the broader concept of force is challenging, even at the high school level (AAAS 1993).

• Some students believe that if a body is not moving, there is no force acting on it (AAAS 1993).

• Students have difficulty understanding that all interactions involve equal forces acting in opposite directions on the separate, interacting bodies. They tend to believe that animate objects (like a person's hands) can exert forces whereas inanimate objects (like tables) cannot (AAAS 1993).

• Some research has shown that teaching high school students to seek consistent explanations for why objects are at rest can help them understand that both "active" and "passive" objects exert forces. Showing students that apparently rigid or supporting objects actually deform might also help them to understand the at-rest condition (AAAS 1993).

• Elementary school students typically do

★ Indicates a strong match between the ideas elicited by the probe and a national standard's learning goal.

not understand gravity as a force. If students do view weight as a force, they often think it is the air that exerts this downward force (AAAS 1993).

- The way children think about forces is related to their meaning for the word *force*. Some students associate force with coercion, physical activity, muscular strength, or living things (Driver et al. 1994).

- Students generally appear to think of force as a property of a single object rather than as a feature of interaction between two objects (Driver et al. 1994).

- Using the example of a book on a table, many students, including high school students, merely think of the table as being in the way, rather than exerting a force (Driver et al. 1994).

Suggestions for Instruction and Assessment

- Provide students with a sequence of scenarios that demonstrate that surfaces deform in a springlike fashion when objects are placed on them and that the tendency of surfaces to return to their original shape causes them to exert force on the object. This type of "bridging analogy" is especially effective with high school students when begun by showing a single spring pushing up on an object, then extended to gradually stiffer objects. By placing a heavy book on this progression of objects, show students how a bedspring, a sponge, and a pair of metersticks spanning a gap between tables all deform in a springlike manner to

exert an upward force on the book (Clement 1993; Minstrell 1982).

- Use examples in instruction and assessment that depict inanimate objects exerting forces on other objects, rather than only involving humans or vehicles as agents of force. Students tend to equate forces with animate objects like people, rolling balls, and wheeled vehicles and thus need multiple opportunities to associate forces with inanimate objects, such as a book resting on a table or a box on the floor.

- The difficulty with younger children is convincing them that any forces at all are acting on a stationary object. Ask them, What would happen if the desk did not support the apple? If they reply that it would fall, ask them, What prevents it from falling? To conceptualize the idea that a force must be exerted, take turns placing a heavy object, such as a brick, in students' hands. Ask the students what they need to do to keep the brick steady and not moving. They should notice that they have to press upward on the brick. Explain how this opposes the downward force of gravity. The downward force and the upward push of their hands are balanced when the brick does not move. If they push harder or relax their muscles, the forces will be unbalanced and the brick will move downward or upward.

- Describing force as an interaction between a pair of objects is often difficult for students to grasp. Encourage students to identify all forces as interactions by naming them as the force of one object on another. This method of describing forces is obvious to students when they refer to examples such as "Shawna's push on Jason," but it is much less obvious and more powerful when students must refer to forces such as "Earth's pull on the apple," or "the table's push on the apple." Asking what is pushing and what is pulling should include mixed examples of cases in which there is motion and cases in which there is no motion.

- In later grades, apply force diagrams frequently, but carefully, in relation to these topics. These diagrams are a vital tool for both learning and assessment, but they can also reinforce student misconceptions if they do not clearly isolate one object at a time and depict the forces acting solely on that object. Such diagrams reveal especially rich information about student thinking when students label each force with the objects involved in the interaction and indicate the relative strengths of forces with arrow lengths (Arons 1997).

- Because student beliefs about the inability of inanimate objects to exert forces can be strikingly firm, simply telling students that this is not so stands a good chance of leaving these students with misconceptions. Instead, work to assess and reflect students' ideas through probing questions and discussion, being careful not to evaluate them. Highlighting ideas of theirs that may contradict visible evidence or other ideas through this type of careful reflective

discussion stands a much greater chance of helping students change their initial ideas.

Related NSTA Science Store Publications and NSTA Journal Articles

American Association for the Advancement of Science (AAAS). 1993. *Benchmarks for science literacy.* New York: Oxford University Press.

American Association for the Advancement of Science (AAAS). 2001. *Atlas of science literacy.* Vol. 1, "laws of motion map," 62–63. Washington, DC: AAAS.

Minstrell, J., and E. van Zee. 2003. Using student questioning to assess and foster student thinking. In *Everyday Assessment in the Science Classroom,* eds. J. M. Atkins, and J. E. Coffey, 61–73. Arlington, VA: NSTA Press.

Robertson, W. 2002. *Force and motion: Stop Faking It! Finally Understanding Science So You Can Teach It.* Arlington, VA: NSTA Press.

Related Curriculum Topic Study Guide
(Keeley 2005)
"Forces"

References

American Association for the Advancement of Science (AAAS). 1993. *Benchmarks for science literacy.* New York: Oxford University Press.

Arons, A. A. 1997. *Teaching introductory physics.* New York: John Wiley and Sons.

Clement, J. 1993. Using bridging analogies and anchoring intuitions to deal with students' pre-

conceptions in physics. *Journal of Research in Science Teaching* 30 (1): 1241–1257.

Driver, R., A. Squires, P. Rushworth, and V. Wood-Robinson. 1994. *Making sense of secondary science: Research into children's ideas.* London and New York: RoutledgeFalmer.

Keeley, P. 2005. *Science curriculum topic study: Bridging the gap between standards and practice.* Thousand Oaks, CA: Corwin Press.

Minstrell, J. 1982. Explaining the "at rest" condition of an object. *The Physics Teacher* 20 (1): 10–14.

National Research Council (NRC). 1996. *National science education standards.* Washington, DC: National Academy Press.

Rolling Marbles

Five friends built a marble tower. The marble tower had a curved track. The track was designed so that the marbles would move down the track in a circular path. The track ended on the floor. Each friend predicted how he or she thought the marble would move when it rolled off the end of the track onto the floor. This is what they said:

Magda: "I think it will roll in circles."

Soledad: "I think it will curve for a bit and then straighten out."

Allen: "I think it will roll in one big curve."

Keira: "I think it will roll in a straight line."

Rafael: "I think it will zigzag for a little while."

Which friend do you most agree with? _____

Use the picture above to draw the path you think the marble will take when it gets to the end of the track.

Explain your thinking. Why do you think the marble will move that way?

_____

_____

_____

_____

_____

_____

_____

_____

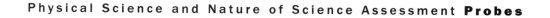

Rolling Marbles

Teacher Notes

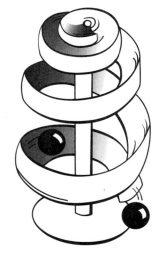

Purpose

The purpose of this assessment probe is to elicit students' ideas about circular motion. The probe is designed to determine whether students recognize that an object will move in a straight line unless acted upon by an outside force.

Related Concepts

circular motion, force, inertia, Newton's first law

Explanation

The best answer is Keira's: The marble leaving the track will travel in a straight line. This behavior is true of all objects: If no outside forces act on an object, the object will travel in a straight line at a constant speed. As the marble rolls down the marble tower's spiral track, a force toward the center of the spiral (a centripetal force) caused by the outside wall keeps the marble rolling in a spiral path. When the marble leaves the end of the track, it is no longer in contact with the walls of the track. Without the track pushing on it, the marble no longer has a center-directed force acting on it that causes it to roll in a curved path. According to Newton's first law, an object will remain at rest or in uniform motion in a straight line unless acted upon by an external force. Since there is no longer a center-directed, external force exerted by the walls of the circular track pushing on the marble, the marble rolls off the end of the track in a straight-line path across the floor. It will continue to do so until an outside force causes it either to change direction or slow down and stop.

Curricular and Instructional Considerations

Elementary Students

Students at the elementary school level may have played with curved marble towers and winding chutes or toy cars moving down curved tracks. Even though their observations may show that a moving object leaves a curved path in a straight line, students tend to revert to their intuitions that an object will continue to move in a curved path. At this level, their experiences are observational, forming a foundation to later develop explanations in middle school that are based on Newton's laws.

Middle School Students

Force and motion relationships are developed more fully at the middle school level. Students learn about Newton's first law of motion and a variety of phenomena that can be explained by it. The idea of inertia is conceptually developed at this level. Students have various experiences observing and explaining a variety of motions, including circular motion.

High School Students

Students build on their previous experience with Newton's first law at the high school level, adding mathematical relationships to their understandings. As they move from qualitative to quantitative views of forces and motion, students begin to understand the mathematical consequences that lead to Newton's laws of motion. Students are now able to use Newton's second law to solve circular motion scenarios from new perspectives and make quantitative predictions with confidence. However, this ability to perform mathematical operations should not be overemphasized in place of conceptual understanding. Students still strongly retain many incorrect ideas about circular motion, and being able to perform mathematical calculations or restate Newton's laws from memory is not a certain indication of understanding.

Administering the Probe

Make sure that students understand what is meant by "a spirally curving marble track." Consider bringing in a prop, such as a child's curved marble tower or a winding car track, or curved tube, such as a hose or flexible pipe, so that students understand the context of the probe. The prop can then be used to test students' predictions after they commit to an outcome.

Related Ideas in *National Science Education Standards* (NRC 1996)

K–4 Position and Motion of Objects

- The position and motion of objects can be changed by pushing or pulling. The size of the change is related to the strength of the push or pull.

5–8 Motions and Forces

- ★ The motion of an object can be described by the object's position, direction, and speed.
- ★ Unbalanced forces will cause changes in the speed or direction of an object's motion.

★ Indicates a strong match between the ideas elicited by the probe and a national standard's learning goal.

9–12 Motions and Forces

★ Objects change their motion only when a net force is applied. Laws of motion are used to calculate precisely the effects of forces on the motion of objects. The magnitude of the change in motion can be calculated using the relationship Force = Mass × Acceleration (F = ma), which is independent of the nature of the force. Whenever one object exerts force on another, a force equal in magnitude and opposite in direction is exerted on the first object.

Related Ideas in *Benchmarks for Science Literacy* (AAAS 1993)

K–2 Motion

• Things move in many different ways, such as straight, zigzag, round and round, back and forth, and fast and slow.

• The way to change how something is moving is to give it a push or a pull.

3–5 Motion

• Changes in speed or direction of motion are caused by forces.

6–8 Motion

★ An unbalanced force acting on an object changes its speed or direction of motion, or both. If the force acts toward a single center, the object's path may curve into an orbit around the center.

9–12 Motion

• The change in motion of an object is proportional to the applied force and inversely proportional to the mass.

Related Research

• Students often expect that objects moving in a curved path because of a wall or constraint will continue to do so when the wall or constraint is removed. This belief that the wall or constraint "trains" the object to follow a curved path is deeply rooted in students and persists even with targeted instruction (Arons 1997).

• Students have difficulty perceiving the direction of motion in a straight line when they encounter situations like an object set in motion inside a curved hollow tube. In this case many students, including those in high school, think the object continues to travel in a curved path when it comes out of the tube (Gunstone and Watts 1985).

• Students' experiences with whirling objects on a string may contribute to their confusion about the direction of the force the string is exerting on the object. Students may think that they are exerting force along the circular path of the object's motion, rather than perpendicular to it, toward the center (Arons 1997).

• Many students think that objects in circular motion are being "thrown outward." This is likely because of the sensation that they feel when traveling around curves in vehicles themselves (Arons 1997, p. 121).

★ Indicates a strong match between the ideas elicited by the probe and a national standard's learning goal.

Suggestions for Instruction and Assessment

- Encourage students to investigate the phenomenon described in the probe themselves, or help them to do so with a demonstration setup. Students who do so should predict the path of the ball's motion and discuss their predictions before performing or viewing the experiment. Several different arrangements can help with this. Try removing a section of an embroidery hoop and rolling a marble along its inside edge, or twirling a soft ball on a string and viewing its motion after the ball is released. Especially with a ball on a string, the motion can be difficult to view, so students should take turns or use a video camera to record and review the motion (Arons 1997, p. 120).

- Ask students about their experiences with marble towers—many are likely to own or have seen a similar setup. Some tracks exist that are flexible enough to reproduce the scenario portrayed in the probe, so students can predict and see the results of the probe itself. Show students the setup before administering the probe, and perform the experiment after discussing students' ideas. Provide an opportunity for students to revise their explanations after viewing the results.

- Talk with students about traveling in a car, on a merry-go-round, or on other amusement park rides. Discuss the feelings they have and relate these to the objects exerting force on them. Use examples from straight-line motion to help students see that, just as they are pushed forward (rather than thrown backward) when a vehicle accelerates, they are *pulled inward* by the vehicle's wall when going around a corner rather than thrown outward.

- If a playground merry-go-round is available, have a student riding on the merry-go-round release a ball after reaching a certain point. Have other students note the motion of the ball: Does it travel in a straight or curved line?

- Attach a string to the side of a wind-up toy that travels in a straight line. Show students that pulling on the string pulls the toy only sideways, and have them pull on the string and watch the toy travel in a curved path. Then have them release the string and watch the motion of the toy as they do so. After they do this several times in a row, have students compare this slow-motion example and the forces acting on the toy to the motion in previous examples that occur more quickly.

- If students have had experiences with hoses, ask them what direction the water flows in when they turn on a spigot and the water comes out of a coiled-up hose. Many students have seen the water shoot out in a straight line, rather than a curved path. Since this phenomenon involves a liquid rather than a solid, make sure students understand that Newton's first law applies to liquids as well as solids.

Related NSTA Science Store Publications and NSTA Journal Articles

American Association for the Advancement of Science (AAAS). 1993. *Benchmarks for science literacy.* New York: Oxford University Press.

American Association for the Advancement of Science (AAAS). 2001. *Atlas of science literacy.* Vol. 1, "laws of motion map," 62–63. New York: Oxford University Press.

Keeley, P. 2005. *Science curriculum topic study: Bridging the gap between standards and practice.* Thousand Oaks, CA: Corwin Press.

National Research Council (NRC). 1996. *National science education standards.* Washington, DC: National Academy Press.

Robertson, W. 2002. *Force and motion, Stop Faking It! Finally Understanding Science So You Can Teach It.* Arlington, VA: NSTA Press.

Stein, M. 1998. Toying with science. *Science and Children* (Sept.): 35–39.

Harris, J. 2004. Science 101: Are there different types of force and motion? *Science and Children* (Mar.): 19.

Related Curriculum Topic Study Guides

(Keeley 2005)
"Forces"
"Laws of Motion"

References

American Association for the Advancement of Science (AAAS). 1993. *Benchmarks for science literacy.* New York: Oxford University Press.

Arons, A. A. 1997. *Teaching introductory physics.* New York: John Wiley and Sons.

Gunstone, R., and M. Watts. 1985. Force and motion. In *Children's Ideas in Science,* eds. R. Driver, E. Guesne, and A. Tiberghien. Milton Keynes, UK: Open University Press.

Keeley, P. 2005. *Science curriculum topic study: Bridging the gap between standards and practice.* Thousand Oaks, CA: Corwin Press.

National Research Council (NRC). 1996. *National science education standards.* Washington, DC: National Academy Press.

Dropping Balls

Reggie has three different types of balls. Each ball is about the same size.

Ball 1 is a wooden ball. Its mass is 28 g.

Ball 2 is a golf ball. Its mass is 46 g.

Ball 3 is a metal ball. Its mass is 110 g.

Reggie held his arm out and dropped the three balls at the same time from the same height. In what order will the balls hit the floor? Circle your prediction:

Prediction A: Ball 1, then ball 2, then ball 3.

Prediction B: Ball 3, then ball 2, then ball 1.

Prediction C: Ball 2, then ball 3, then ball 1.

Prediction D: All three balls will hit the floor at about the same time.

Prediction E: Ball 3 will hit first, followed by ball 1 and ball 2 hitting the floor at the same time.

Explain your thinking. What "rule" or reasoning did you use to make your prediction?

_____

_____

_____

_____

_____

_____

_____

Dropping Balls

Teacher Notes

Purpose

The purpose of this assessment probe is to elicit students' ideas about falling objects. The probe is designed to find out if students think the weight or mass of an object affects how fast it falls.

Related Concepts

acceleration, force, gravity, mass

Explanation

Prediction D is the best answer: All three balls will hit the floor at about the same time. The word *about* is used because the balls are not dropped in a perfect vacuum. In general, freely falling objects increase their speed at the same rate, 9.8 m/s each second, regardless of their mass. Ball 3 has more mass than the other two balls. Because of this, some students reason

that if all three balls are dropped at the same time, the heaviest ball (ball 3) should reach the ground first. This is a seemingly logical conclusion, but it actually hits the floor at the same time as the lighter balls due to the way forces affect the motion of objects.

Objects with more mass accelerate more slowly than objects with less mass, if pushed or pulled with equal force. This phenomenon is clear to anyone who has pushed a very heavy object in order to start it moving. If you could push an object with the same constant force every time, you would actually find that an object with *twice* the mass speeds up exactly *half* as quickly. On the other hand, if you pushed twice as hard on any given object, the object would speed up twice as quickly.

This same effect applies to falling objects like balls. At first thought, it would seem that

the stronger gravitational force on the metal ball should speed it up more quickly. In fact, the stronger pull on this ball is canceled out by the extra effort required to speed up this more massive ball. If Reggie decided to drop a ball that had five times more mass than ball 3, the force pulling the ball downward would be five times greater, but the ball would also be five times more difficult to speed up. As a result, this very heavy ball would speed up equally quickly and hit the ground at the same time as ball 1, 2, or 3. The rate at which free-falling objects fall is 9.8 m/s/s, regardless of their mass.

However, because the balls are falling through the air, the air exerts an upward force on them as they fall. This force can differ depending on an object's size, density, and speed. Because of this force, objects dropped together in an experiment will not hit the ground at the same time if air resistance is a factor. For example, if a Ping-Pong ball with a mass of 2.7 g were dropped with the three balls in the probe, it would not hit the ground at the same time as the other three because air resistance would be a significant factor. If a sheet of paper and a book were dropped at the same time from the same height, the book would land first because air resistance cannot be ignored with the sheet of paper. This probe assumes that the effect of air resistance on the balls is negligible. If the air could be removed and the balls dropped in a vacuum, they would always hit the ground at exactly the same time.

Curricular and Instructional Considerations

Elementary Students

In the elementary school grades, students observe that things fall when dropped and later relate this to Earth's gravitational pull. They observe how different objects fall and that they can change how fast an object falls by changing its shape. At this stage, their investigations of the motion of falling objects are primarily observational.

Middle School Students

In the middle school grades, students engage in concrete experiences involving force and motion from which a more comprehensive understanding of force and motion can be developed later in high school. Students observe the effects of different forces on falling, rolling, and sliding objects and begin to move from qualitative descriptions to quantitative ones.

High School Students

Students at the high school level move from understanding gravity as a general universal force to understanding more of the details and mathematical description of gravitational forces. At this level, students are better able to engage in the more abstract thinking involved with mathematical representations, such as the acceleration of a falling object, and also to learn of the many contexts in which gravity plays an important role. High school students are also able to move from qualitative descriptions of motion toward quantitative ones and should

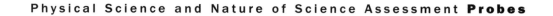

be encouraged to do so (AAAS 1993). Before launching into applications of mathematical formulae, students should demonstrate the ability to describe the subtleties involved in changing motion. By learning mathematical descriptions and specific terminology associated with motion concepts (*velocity, acceleration,* etc.) and after acquiring a broad foundation of firsthand experiences observing and describing motion, students can use vocabulary and calculations with meaning. At this grade level the concept of force is better understood but students' intuitive ideas about the effects of forces on objects are still tenacious.

Administering the Probe

Show students a wooden ball, golf ball, and similarly sized metal ball and have them hold them to feel their "felt weight" (or show three other similarly shaped objects of different masses). Make sure that students know that the balls will be released at exactly the same time, from exactly the same height.

Related Ideas in *National Science Education Standards* (NRC 1996)

K–4 Position and Motion of Objects

*   The position and motion of objects can be changed by pushing or pulling. The size of the change is related to the strength of the push or pull.

5–8 Motions and Forces

★ If more than one force acts on an object along a straight line, then the forces will reinforce or cancel one another, depending on their direction and magnitude. Unbalanced forces will cause changes in the speed or direction of an object's motion.

9–12 Motions and Forces

*   Objects change their motion only when a net force is applied. Laws of motion are used to calculate precisely the effects of forces on the motion of objects. The magnitude of the change in motion can be calculated using the relationship Force = Mass × Acceleration, which is independent of the nature of the force.

*   Gravitation is a universal force that each mass exerts on any other mass.

Related Ideas in *Benchmarks for Science Literacy* (AAAS 1993)

K–2 Motion

*   Things near the Earth fall to the ground unless something holds them up.

*   Things move in many different ways, such as straight, zigzag, round and round, back and forth, and fast and slow.

3–5 Motion

*   Changes in speed or direction of motion are caused by forces. The greater the force is, the greater the change in motion will be.

*   The Earth's gravity pulls any object toward it without touching it.

★ Indicates a strong match between the ideas elicited by the probe and a national standard's learning goal.

6–8 Motion

- An unbalanced force acting on an object changes its speed or direction of motion, or both.
- Every object exerts gravitational force on every other object.

9–12 Motion

- The change in motion of an object is proportional to the applied force and inversely proportional to the mass.

Related Research

- Students do not always identify a force to account for falling objects. They think objects "just fall naturally" or that the person letting go of the object has caused it to fall (Driver et al. 1994).
- Studies by Osborne (1984) found that students think heavier objects fall faster.
- Students, including university students, tend to think that heavier objects fall to Earth faster because they have a bigger acceleration due to gravity (Driver et al. 1994).

Suggestions for Instruction and Assessment

- This probe lends itself well to an inquiry investigation. Have students try this with the materials mentioned in the probe or with other objects such as different size rocks, different coins, or blocks made of different materials. Realize, however, that dropping two or more objects in exactly the same way and at the same time is difficult, and that this and some effects of air

resistance can make it hard to reproduce the expected results. As a result, student investigations alone can run the risk of reinforcing incorrect student ideas. You may want to help guide students toward procedures, including dropping only two objects at a time, and materials that minimize these problems.

- Encourage students to investigate changing how some objects fall in order to observe the effects of air resistance. Have students drop a piece of paper both unfolded and as a crumpled ball. Also have students drop paper and a book side-by-side and then with the paper placed on top of the book. As with most experiences, students should make predictions about what they expect to see prior to these experiences and discuss the results afterward.

- Show students videos of this situation performed under highly controlled conditions—many examples can be found. A film of an astronaut dropping a hammer and feather together on the Moon is especially interesting. However, be aware this film may cause students to think that something special about being on the Moon, such as less gravity, causes the objects to drop together. A movie of the Apollo 14 "Hammer-Feather Drop" can be viewed online at *http://nssdc.gsfc.nasa.gov/planetary/lunar/apollo_15_feather_drop.html*.

- Connect this probe scenario to the historical example of Galileo's famous experiments with falling objects.

Related NSTA Science Store Publications and NSTA Journal Articles

American Association for the Advancement of Science (AAAS). 1993. *Benchmarks for science literacy.* New York: Oxford University Press.

American Association for the Advancement of Science (AAAS). 2001. *Atlas of science literacy.* Vol. 1, "laws of motion," 62–63. New York: Oxford University Press.

Driver, R., A. Squires, P. Rushworth, and V. Wood-Robinson. 1994. *Making sense of secondary science: Research into children's ideas.* London and New York: RoutledgeFalmer.

Keeley, P. 2005. *Science curriculum topic study: Bridging the gap between standards and practice.* Thousand Oaks, CA: Corwin Press.

National Research Council (NRC). 1996. *National science education standards.* Washington, DC: National Academy Press.

Nelson, G. 2004. Science 101: What is gravity? *Science and Children* (Sep.): 22–23.

Robertson, W. 2002. *Force and motion, Stop Faking It! Finally Understanding Science So You Can Teach It.* Arlington, VA: NSTA Press.

Related Curriculum Topic Study Guides

(Keeley 2005)

"Forces"

"Gravity"

"Motion"

References

American Association for the Advancement of Science (AAAS). 1993. *Benchmarks for science literacy.* New York: Oxford University Press.

Driver, R., A. Squires, P. Rushworth, and V. Wood-Robinson. 1994. *Making sense of secondary science: Research into children's ideas.* London and New York: RoutledgeFalmer.

Keeley, P. 2005. *Science curriculum topic study: Bridging the gap between standards and practice.* Thousand Oaks, CA: Corwin Press.

National Research Council (NRC). 1996. *National science education standards.* Washington, DC: National Academy Press.

Osborne, R. 1984. Children's dynamics. *The Physics Teacher* 22 (8): 504–508.

Is It a Theory?

Put an X next to the statements you think best apply to scientific theories.

_____ **A**   Theories include observations.

_____ **B**   Theories are "hunches" scientists have.

_____ **C**   Theories can include personal beliefs or opinions.

_____ **D**   Theories have been tested many times.

_____ **E**   Theories are incomplete, temporary ideas.

_____ **F**   A theory never changes.

_____ **G**   Theories are inferred explanations, strongly supported by evidence.

_____ **H**   A scientific law has been proven and a theory has not.

_____ **I**   Theories are used to make predictions.

_____ **J**   Laws are more important to science than theories.

Examine the statements you checked off. Describe what a theory in science means to you.

_____

_____

_____

_____

_____

_____

_____

Is It a Theory?

Teacher Notes

Purpose

The purpose of this assessment probe is to elicit students' ideas about the nature of science. The probe is designed to find out if students distinguish scientific theories from the common use of the word *theory* and if they understand how theories differ from laws.

Related Concepts

hypothesis, nature of science, scientific law, theory

Explanation

The statements that best describe scientific theories are A, D, G, and I. Scientific theories are evidence-based explanations based on related observations of phenomena or events. A scientific theory is based on a solid body of sup-

porting evidence that has been tested and supported with multiple lines of evidence. Theories are widely accepted in the scientific community and can be used to make predictions. Theories in science are not kept in doubt, although because of the dynamic nature of science, they can change if new evidence becomes available. Such new evidence may be made possible through new technological tools, techniques of analysis, new theoretical advances, or shifts in research emphasis that lead the scientific community to reconsider an existing explanation and revise it to fit new evidence that is available and accepted. Theories can also change when scientists view the same evidence differently, such as the example of Darwinian evolution and punctuate evolution in which the same evidence was looked at from a different perspective.

Examples of scientific theories include the germ theory of disease, the theory of biological evolution, plate tectonics theory, string theory, big bang theory, and kinetic molecular theory. Each of these theories provide an explanation accepted by the scientific community for observed phenomena. For example, plate tectonics explains the observed evidence for large-scale motions of the Earth's lithosphere.

Students and nonscientist adults often have definitions for the word *theory* that are quite different from the scientific meaning of the word. To nonscientists, the word *theory* often means a hunch, opinion, or a guess. In common usage it is not unusual to hear someone say, "I have a theory about…." A theory in the nonscientific sense of the word does not require firm evidence to support it nor does it require the consensus of others.

Sometimes the words *hypothesis, theory,* and *law* are inaccurately portrayed in science textbooks as an "evolution" of a scientific idea. There isn't a definite sequence or hierarchy for the development of scientific ideas—such as a hypothesis leads to a theory, which eventually becomes a law—because they represent different types of knowledge. For example, it is possible to develop a law (observed behavior of nature) and not have the explanation (theory) for it, such as when Isaac Newton helped develop the law of gravity, but at the time he did not have an explanation for it.

Law and theory are two different key elements of the nature of scientific knowledge. Laws are generalizations, principles, or patterns in nature derived from scientific facts that often describe how the natural world behaves under certain conditions. Laws describe relationships among observable phenomena. Some laws are expressed mathematically. Examples of scientific laws include Newton's laws of motion, universal law of gravitation, Boyle's law, and Mendel's laws. A law describes a phenomenon or event but it does not explain it, like a theory does. A theory is not a "law in waiting." Theories do not mature into laws (Lederman and Lederman 2004). A theory is a well-established explanation. Laws describe *what,* and theories explain *why.*

Curricular and Instructional Considerations

Elementary Students

From their very first day in school, young students should be actively engaged in learning to view the world scientifically. They should be encouraged to ask questions about nature and to seek answers, collect things, count and measure things, make qualitative observations, organize collections and observations, discuss findings, and so on. These skills and activities are precursors to understanding how science relies on evidence. Getting into the spirit of science and enjoying science are important at this age. Students can learn some things about the nature of scientific inquiry and significant people from history, which will provide a foundation for the development of sophisticated ideas related to the history and nature of science that will be developed in later years (NRC 1996).

Middle School Students

In middle school, students begin to deal with the changing nature of scientific knowledge. Both incremental changes and more radical changes in scientific knowledge should be taken up (AAAS 1993). At this grade level, students should be introduced to the scientific meaning of the word *theory* and become familiar with scientific theories (and their historical development) that are appropriately connected to the content they are learning (e.g., germ theory).

High School Students

Students at this level should be able to distinguish between facts, hypotheses, theories, and laws. Their formal understanding of what a scientific theory is should be developed both through historical episodes in science and by reflecting on developments in current science. This is a time when it is important to precede the teaching of important theories that are central to the different disciplines of science, such as the centrality of the theory of evolution in biology, with explicit teaching of the nature of science. For example, biological evolution is one of the strongest, most important, and useful scientific theories we have in science (NRC 1996). By helping students understand what a theory is in science before teaching about major theories such as biological evolution, teachers are also helping students to better understand why a theory is accepted by the scientific community and how personal beliefs and religious views that are not based on scientific evidence are not part of learning science.

Administering the Probe

This probe is best used at the middle school and high school levels. It can be used as a paper-pencil assessment to gather students' ideas for later analysis as well as a stimulus for provoking discussion about the nature of science. It can also be administered as a card sort with small groups of students sorting each statement into two groups, "applies to scientific theories" or "does not apply to scientific theories," while defending their reasons for placing each card.

Related Ideas in *National Science Education Standards* (NRC 1996)

. . . .

5–8 Understandings About Scientific Inquiry

★ Scientific explanations emphasize evidence, have logically consistent arguments, and use scientific principles, models, and theories. The scientific community accepts and uses such explanations until displaced by better scientific ones. When such displacement occurs, science advances.

5–8 The History and Nature of Science

★ Scientists formulate and test their explanations of nature using observation, experiments, and theoretical and mathematical models. Although all scientific ideas are tentative and subject to change and improvement in principle, for most major ideas in science, there is much experimental and observational confirmation. Those ideas are not likely to change greatly in

---

★ Indicates a strong match between the ideas elicited by the probe and a national standard's learning goal.

the future. Scientists do change and have changed their ideas about nature when they encounter new experimental evidence that does not match their existing explanations.

- In areas where active research is being pursued and in which there is not a great deal of experimental or observational evidence and understanding, it is normal for scientists to differ with one another about the interpretation of the evidence or theory being considered.

9–12 The History and Nature of Science

- Science distinguishes itself from other ways of knowing and from other bodies of knowledge through the use of empirical standards, logical arguments, and skepticism as scientists strive for the best possible explanations about the natural world.

- ★ Scientific explanations must meet certain criteria. First and foremost, they must be consistent with experimental and observational evidence about nature and must make accurate predictions, when appropriate, about systems being studied. Scientific explanations should also be logical, respect the rules of evidence, be open to criticism, report methods and procedures, and make knowledge public. Explanations on how the natural world changes based on myths, personal beliefs, religious values, mystical inspiration, superstition, or authority may be personally useful and socially relevant, but they are not scientific.

- Because all scientific ideas depend on exper-

imental and observational confirmation, all scientific knowledge is, in principle, subject to change as new evidence becomes available. The core ideas of science, such as the conservation of energy or the laws of motion, have been subjected to a wide variety of confirmations and are therefore unlikely to change in the areas in which they have been tested. In areas where data or understanding are incomplete, such as the details of human evolution or questions surrounding global warming, new data may well lead to changes in current ideas or resolve current conflicts. In situations where information is still fragmentary, it is normal for scientific ideas to be incomplete, but this is also where the opportunity for making advances may be greatest.

Related Ideas in *Benchmarks for Science Literacy* (AAAS 1993)

6–8 The Scientific World View

- ★ Scientific knowledge is subject to modification as new information challenges prevailing theories and as a new theory leads to looking at old observations in a new way.

- Some matters cannot be examined usefully in a scientific way. Among them are matters that by their nature cannot be tested objectively and those that are essentially matters of morality.

9–12 The Scientific World View

- From time to time, major shifts occur in

★ Indicates a strong match between the ideas elicited by the probe and a national standard's learning goal.

the scientific view of how the world works. More often, however, the changes that take place in the body of scientific knowledge are small modifications of prior knowledge. Change and continuity are persistent features of science.

★ No matter how well one theory fits a set of observations, a new theory might fit it just as well or better, or it might fit a wider range of observations. In science, the testing, revising, and occasional discarding of theories, new and old, never ends. This ongoing process leads to an increasingly better understanding of how things work in the world but not to absolute truth. Evidence for the value of this approach is given by the improving ability of scientists to offer reliable explanations and make accurate predictions.

9–12 Scientific Inquiry

★ In the short run, new ideas that do not mesh well with mainstream ideas in science often encounter vigorous criticism. In the long run, theories are judged by how they fit with other theories, the range of observations they explain, how well they explain observations, and how effective they are in predicting new findings.

• New ideas in science are limited by the context in which they are conceived; are often rejected by the scientific establishment; sometimes spring from unexpected findings; and usually grow slowly, through contributions from many investigators.

Related Research

• Students of all ages find it difficult to distinguish between a theory and the evidence for it, or between description of evidence and interpretation of evidence (AAAS 1993).

• Young students often state that a theory involves knowing something about the situation, but they offer no further elaboration (Driver et al. 1996).

• Students often use *theory* to describe a prediction, as in "I have a theory about how that works," that is based on a guess rather than evidence (Driver et al. 1996).

• Middle school students have difficulty understanding the development of scientific knowledge through the interaction of theory and observation (AAAS 1993).

• Younger students tend to characterize testing as a simple process of observation with the outcomes being obvious, while older students seem to be more aware that testing may involve finding out about mechanisms or testing theories (Driver et al. 1996).

• Studies have indicated that students' understanding of evolution is related to their understanding of the nature of science, including understanding what constitutes a theory and their general reasoning abilities (AAAS 1993).

• Over the past two decades, there has been a considerable awareness and acceptance of the importance of developing an understanding of the nature of science among students and among teachers. Research shows that a deep, conceptual understand-

---

★ Indicates a strong match between the ideas elicited by the probe and a national standard's learning goal.

ing of the nature of science has not been attained (Lederman et al. 2002), despite attempts to improve both students' and teachers' views of the nature of science.

Suggestions for Instruction and Assessment

- Teaching about the nature of science can get lost if it is embedded within regular science instruction (Crowther, Lederman, and Lederman 2005). Nature of science can be embedded within traditional content but should be explicitly and formally taught and assessed as a subject matter topic in much the same way that curricular topics such as life cycles, chemical reactions, or phases of the Moon are taught (Abd-El-Khalick, Bell, and Lederman 1998).

- Almost any science activity can be modified to explicitly teach some aspect of nature of science. An NSTA article by Lederman and Lederman (2004) provides an example of how a typical science activity, such as observing the stages of mitosis, can embed lessons on the nature of science.

- Use historical events to help middle and high school students develop an understanding of the nature of science and development of theories. *Atlas of Science Literacy,* Volume 2 (AAAS 2007), makes key curricular connections between the nature of science and the development of historical ideas, such as Copernican theory, theory of plate tectonics, Einstein's theory of relativity, and more.

- Make teaching the nature of science, in-

cluding understanding what a theory is, an integral component of the middle school and high school science curriculum and a conscious, deliberate part of teaching. Do not assume students pick up these formal understandings of science automatically by engaging in inquiry-based activities. They should be explicitly taught and formally included in the science curriculum throughout the year, not just as the typical introductory chapter in a textbook.

- Today's students are tomorrow's teachers, parents, politicians, and world leaders. It is essential that today's students be explicitly guided to develop the skills and knowledge to distinguish scientific knowledge from religious, cultural, philosophical, or other beliefs that are not grounded in scientific evidence and can impact the way people address important questions and make informed decisions that affect their lives, society, and the natural world. Teachers should pay careful attention to developing the knowledge students need to hone their ability to distinguish science-based information from that which is not grounded in scientific thinking so that they can make informed decisions about matters that often end up in the public arena.

- A technique to help students maintain a consistent image of the nature of science throughout the year by paying more careful attention to the words they use is to create a "caution words" poster or bulletin board (Schwartz 2007). Important words that have specific meanings in science but are

often used inappropriately in the science classroom and through everyday language can be posted in the room as a reminder to pay careful attention to how students are using these words. For example, words like *hypothesis, theory,* and *law* can be included on the list.

- Activities with pattern cubes provide students with a common experience to develop an understanding of the nature of science and the appropriate language used to describe how science is conducted. Pattern cubes and a description of the activity are available online at *www.nap.edu/readingroom/books/evolution98/evol6-a.html.*

- Take the time during lessons to have discussions with students about the nature of science and encourage reflection on the way they view scientists, scientific knowledge, and scientific practices.

- There are several excellent, diagnostic questionnaires developed by Lederman et al. (2002) that can be used to find out students' views about the nature of science. An internet search can help you locate a source of these questionnaires.

Related NSTA Science Store Publications and NSTA Journal Articles

American Association for the Advancement of Science (AAAS). 1993. *Benchmarks for science literacy.* New York: Oxford University Press.

American Association for the Advancement of Science (AAAS). 2007. *Atlas of science literacy.* Vol. 2. Washington, DC: AAAS.

BSCS. 2005. *The nature of science and the study of biological evolution.* Colorado Springs, CO: BSCS.

Byoung-Sug, K., and M. McKinney. 2007. Teaching the nature of science through the concept of living. *Science Scope* 31 (3): 20–25.

Crowther, D., N. Lederman, and J. Lederman. 2005. Understanding the true meaning of nature of science. *Science and Children* (Oct.): 50–54.

Keeley, P. 2005. *Science curriculum topic study: Bridging the gap between standards and practice.* Thousand Oaks, CA: Corwin Press.

Lederman, N., and J. Lederman. 2004. Revising instruction to teach nature of science. *The Science Teacher* 71 (9): 36–39.

Lederman, N., F. Abd-El-Khalick, and R. Bell. 2000. If we want to talk the talk we must also walk the walk: The nature of science, professional development, and educational reform. In *Issues in science education: Professional development planning and design,* eds. J. Rhoton and P. S. Bowers. Arlington, VA: NSTA Press.

National Research Council (NRC). 1996. *National science education standards.* Washington, DC: National Academy Press.

National Research Council (NRC). 1998. *Teaching about evolution and the nature of science.* Washington, DC: National Academy Press.

National Science Teachers Association (NSTA). 2004. History and nature of science. *The Science Teacher* (Nov.), theme issue.

Schwartz, R. 2007. What's in a word? How word choice can develop (mis)conceptions about the nature of science. *Science Scope* 31 (2): 42–47.

Sweeney-Lederman, J., and N. Lederman. 2005. Science shorts: Nature of science is... *Science and Children* (Oct.): 53–54.

Related Curriculum Topic Study Guide

(Keeley 2005)

"The Nature of Scientific Thought and Development"

References

Abd-El-Khalick, F., R. Bell, and N. Lederman. 1998. The nature of science and instructional practice: Making the unnatural natural. *Science Education* 82: 417–436.

American Association for the Advancement of Science (AAAS). 1993. *Benchmarks for science literacy.* New York: Oxford University Press.

American Association for the Advancement of Science (AAAS). 2007. *Atlas of science literacy.* Vol. 2. Washington, DC: AAAS.

BSCS. 2005. *The nature of science and the study of biological evolution.* Colorado Springs, CO: BSCS.

Crowther, D., N. Lederman, and J. Lederman. 2005. Understanding the true meaning of nature of science. *Science and Children* (Oct.): 50–54.

Driver, R., J. Leach, R. Millar, and P. Scott. 1996. *Young people's images of science.* Buckingham, UK: Open University Press.

Keeley, P. 2005. *Science curriculum topic study: Bridging the gap between standards and practice.* Thousand Oaks, CA: Corwin Press.

Lederman, N., and J. Lederman. 2004. Revising instruction to teach nature of science. *The Science Teacher* 71 (9): 36–39.

Lederman, N., F. Abd-El-Khalick, R. Bell, and R. Schwartz. 2002. Views of nature of science questionnaire: Toward valid and meaningful assessment of learner's conceptions of nature of science. *Journal of Research in Science Teaching* 39 (6): 497–521.

National Research Council (NRC). 1996. *National science education standards.* Washington, DC: National Academy Press.

Schwartz, R. 2007. What's in a word? How word choice can develop (mis)conceptions about the nature of science. *Science Scope* 31 (2): 42–47.

Doing Science

Four students were having a discussion about how scientists do their work. This is what they said:

Antoine: "I think scientists just try out different things until something works."

Tamara: "I think there is a definite set of steps all scientists follow called the scientific method."

Marcos: "I think scientists use different methods depending on their question."

Avery: "I think scientists use different methods but they all involve doing experiments."

Which student do you most agree with? _____

Explain why you agree with that student and include why you disagree with the other students.

_____

_____

_____

_____

_____

_____

_____

_____

_____

_____

_____

_____

_____

Doing Science

Teacher Notes

Purpose

The purpose of this assessment probe is to elicit students' ideas about scientific investigations. The probe is designed to find out if students recognize that scientists investigate the natural world in a variety of ways depending on the question they pose and that there is no fixed sequence of steps called the "scientific method" that all scientists use and follow rigidly.

Related Concepts

experiment, nature of science, scientific inquiry, scientific method

Explanation

The best answer is Marcos's: I think scientists use different methods depending on their question. Doing science is generally a logical, systematic process, unlike Antoine's response,

which implies that the approach to science is random rather than methodical. Sometimes creative, divergent thinking and approaches have led to scientific discovery but they usually involve a systematic approach. Fundamentally, the various scientific disciplines are alike in their reliance on evidence, the use of hypotheses and theories when appropriate, the kinds of logic used, and more; however, scientists differ greatly from one another in what phenomena they investigate and in how they go about their work (AAAS 1988, p. 4). The scientific method correctly implies a methodical approach; however, Tamara's response implies that there is one method that includes a definite sequence of steps that all scientists follow. "There simply is no fixed set of steps that scientists always follow, no one path that leads them unerringly to scientific knowledge" (AAAS 1988, p. 4). Scientists move back and forth among processes and do not follow a recipe.

Experimentation is a process in which scientists control conditions in order to test their hypotheses. Unlike Avery's response, not all scientific investigations involve experiments. An experiment is a type of investigation that involves testing cause-and-effect relationships between variables—manipulated (independent) and responding (dependent). Astronomy, field studies in nature, and paleontology are some of the examples of areas of science in which it would be difficult or unfeasible to manipulate and control experimental conditions. In these types of investigations, scientists rely on a wide range of naturally occurring observations to make inferences about organisms, objects, events, or processes. For example, the link between smoking and lung cancer was actually established through correlational research designs as opposed to classic experiments.

Curricular and Instructional Considerations

Elementary Students

From their very first day in school, young students should be actively engaged in using science to investigate the world around them. They should be encouraged to ask questions about familiar phenomena and objects and to seek answers, collect things, count and measure things, make qualitative observations, organize collections and observations, and discuss findings in a systematic way. These early experiences with inquiry are precursors to un-

derstanding how science is done in a variety of ways and how it relies on gathering data to use for evidence. By directly experiencing a variety of ways that questions can be answered in science through simple investigations, students will begin to develop the idea that there is no one fixed way to go about answering scientific questions. However, this will only happen if students are asked to reflect on what they have done and instruction explicitly addresses the understandings of inquiry. At this level, students develop the notion of a "fair test" when designing experiments, but caution should be used at this early stage to not imply that all scientific investigations are experiments.

Middle School Students

By middle school, students should understand that science is guided by the question posed. The question and the particular content of the inquiry determine the method used to investigate. Caution is taken to ensure that students do not develop the idea that there is one "scientific method" that involves a prescribed set of linear steps that all scientists follow. At this level, students progress beyond the notion of a fair test to include a formal understanding of experimentation as a way of testing ideas that involves identifying and controlling variables. Often, the scientific method is taught and used in the context of doing experiments. However, at this grade level students need to experience and understand that science is systematically carried out in a variety of ways, including doing experiments, but not limited to that. Students should also become aware of how different do-

mains of science use different methodologies (e.g., in astronomy, observations are made using remote technologies).

High School Students

In high school, students develop more sophisticated abilities and understandings of scientific inquiry. They are able to design and carry out more complex experiments as a way to systematically test their ideas. At this level, they should also engage in using a variety of other methods to investigate their questions, including field studies, observations of remote or microscopic phenomena using technology, modeling, specimen collections, and so on. This is the time when students should have opportunities to read and analyze peer-reviewed published scientific papers that show the variety of methodologies scientists use to do their work.

Administering the Probe

This probe is best used as is at the middle school and high school levels, particularly if students have been previously exposed to the term *scientific method* somewhere in their K–12 science education. However, the language of the probe can be modified as a simpler version for K–5 students. Be sure to emphasize that students should explain not only why they agree with the choice they selected from the four responses, but also why they did not select the other choices. The last selected response (Avery) can be expanded to include "…they all involve developing hypotheses and doing experiments."

Related Ideas in *National Science Education Standards* (NRC 1996)

K–4 Understandings About Scientific Inquiry

★ Scientists use different kinds of investigations depending on the questions they are trying to answer. Types of investigations include describing objects, events, and organisms; classifying them; and doing a fair test (experimenting).

5–8 Understandings About Scientific Inquiry

★ Different kinds of questions suggest different kinds of investigations. Some investigations involve observing and describing objects, organisms, or events; some involve collecting specimens; some involve experiments; some involve seeking more information; some involve discovery of new objects and phenomena; and some involve making models.

• Current scientific knowledge and understanding guide scientific investigations. Different scientific domains employ different methods, core theories, and standards to advance scientific knowledge and understanding.

5–8 The History and Nature of Science

• Scientists formulate and test their explanations of nature using observation, experiments, and theoretical and mathematical models.

★ Indicates a strong match between the ideas elicited by the probe and a national standard's learning goal.

9–12 Understandings About Scientific Inquiry

- Scientists usually inquire about how physical, living, or designed systems function. Conceptual principles and knowledge guide scientific inquiries. Historical and current scientific knowledge influence the design and interpretation of investigations and the evaluation of proposed explanations made by other scientists.

Related Ideas in *Benchmarks for Science Literacy* (AAAS 1993)

K–2 Scientific Inquiry

- People can often learn about things around them by just observing those things carefully. Sometimes they can learn more by doing something to the things around them and noting what happens.

3–5 Scientific Inquiry

- ★ Scientific investigations may take many different forms, including observing what things are like or what is happening somewhere, collecting specimens for analysis, and doing experiments.

6–8 Scientific Inquiry

- ★ Scientists differ greatly in what phenomena they study and how they go about their work. Although there is no fixed set of steps that all scientists follow, scientific investigations usually involve the collection of relevant evidence, the use of logical reasoning, and the application of imagination in devising hypotheses and explanations to make sense of the collected evidence.

9–12 Scientific Inquiry

- Investigations are conducted for different reasons, including to explore new phenomena, to check on previous results, to test how well a theory predicts, and to compare different theories.
- ★ Sometimes scientists can control conditions in order to obtain evidence. When that is not possible for practical or ethical reasons, they try to observe as wide a range of natural occurrences as possible to be able to discern patterns.

Related Research

- The idea that there is a common series of steps that is followed by all scientists is likely to be the most common myth of science (McComas 1998).
- Students generally have difficulty with explaining how science is conducted because they have had little contact with real scientists. Their familiarity with doing science, even at older ages, is "school science," which is often not how science is generally conducted in the scientific community (Driver et al. 1996).
- Despite over 10 years of reform efforts in science education, research still shows that students typically have inadequate conceptions of what science is and what scientists do (Schwartz 2007).

★ Indicates a strong match between the ideas elicited by the probe and a national standard's learning goal.

Suggestions for Instruction and Assessment

- The scientific method is often the first topic students encounter when using textbooks, and this can erroneously imply that there is a rigid set of steps all scientists follow. Often the scientific method described in textbooks applies to experimentation, which is only one of many ways scientists conduct their work. Embedding explicit instruction of the various ways to do science in the actual investigations students do throughout the year as well as in their studies of investigations done by scientists is a better approach to understanding how science is done than starting off the year with the so-called scientific method in a way that is devoid of a context through which students can learn the content and process of science.

- Use caution when referring to the scientific method. It may be better to refer to *a* scientific method rather than *the* scientific method in order not to imply that there is one, fixed method.

- Be aware that even though scientists may refer to the term *scientific method,* they use this term to generalize the systematic process of doing science, not a rigid, fixed set of steps all scientists follow.

- Use caution when asking students to write lab reports that use the same format regardless of the type of investigation conducted. The format used in writing about an investigation may imply a rigid, fixed process or may erroneously misrepresent aspects of science, such as the idea that hypotheses are developed for every scientific investigation.

- Be careful how the word *experiment* is used. Students and some teachers use *experiment* as a general term for investigation rather than a specific type of investigation that involves variables or fair testing. Consistently remind students to consider, and explicitly point out, that all experiments are investigations but not all investigations are experiments.

- A technique to help students maintain a consistent image of science as inquiry throughout the year by paying more careful attention to the words they use is to create a "caution words" poster or bulletin board (Schwartz 2007). Important words that have specific meanings in science but are often used inappropriately in the science classroom and through everyday language can be posted in the room as a reminder to pay careful attention to how students are using these words. For example, words like *experiment, hypothesis,* and *scientific method* can be posted as a caution to be careful in their use.

- Opportunities to experience a variety of ways science is conducted is not enough for students to develop deep understandings about inquiry. For students to understand that science can be carried out in a variety of ways, students need to be given time to reflect on what they have done. Like the nature of science, these understandings need to be explicitly addressed.

- Provide students with a variety of ways to investigate scientific questions, including experiments, field observations, modeling, collecting specimens, making remote observations, and so on. Point out how each has its own methodologies depending on the question being asked and the domain of study. For example, an astronomer cannot control conditions in space but can make observations using technology to help understand astronomical phenomena. Point out the similarities of the different ways to do science, including use of existing knowledge, an organized or systematic approach, and reliance on evidence.

- Use historical accounts and nonfiction readings from articles and books about scientists doing their work, such as *The Beak of the Finch* (Weiner 1994).

- Ensuring that students develop the abilities to carry out scientific inquiries is an important part of the standards. However, it is just as important for students to develop understandings of inquiry. Students can perform investigations yet not understand why they are done in a particular way.

Related NSTA Science Store Publications and NSTA Journal Articles

American Association for the Advancement of Science (AAAS). 1993. *Benchmarks for science literacy.* New York: Oxford University Press.

Keeley, P. 2005. *Science curriculum topic study: Bridging the gap between standards and practice.* Thousand Oaks, CA: Corwin Press.

National Research Council (NRC). 1996. *National science education standards.* Washington, DC: National Academy Press.

NSTA Position Statement on Scientific Inquiry. *www.nsta.org/about/positions/inquiry.aspx*

Schwartz, R. 2007. What's in a word? How word choice can develop (mis)conceptions about the nature of science. *Science Scope* 31 (2): 42–47.

Sullivan, M. 2006. *All in a day's work: Careers using science.* Arlington, VA: NSTA Press.

Watson, S., and L. James. 2004. Science sampler: The scientific method—is it still useful? *Science Scope* (Nov./Dec.): 37–39.

Related Curriculum Topic Study Guide

(Keeley 2005)

"Understandings About Scientific Inquiry"

References

American Association for the Advancement of Science (AAAS). 1988. *Science for all Americans.* New York: Oxford University Press.

American Association for the Advancement of Science (AAAS). 1993. *Benchmarks for science literacy.* New York: Oxford University Press.

Driver, R., J. Leach, R. Millar, and P. Scott. 1996. *Young people's images of science.* Buckingham, UK: Open University Press.

Keeley, P. 2005. *Science curriculum topic study: Bridging the gap between standards and practice.* Thousand Oaks, CA: Corwin Press.

McComas, W. 1998. The principle elements of the nature of science: Dispelling the myths. In *The nature of science in science education: Rationales*

and strategies, 53–70. Boston, MA: Kluwer Academic Publishers.

National Research Council (NRC). 1996. *National science education standards.* Washington, DC: National Academy Press.

Schwartz, R. 2007. What's in a word? How word choice can develop (mis)conceptions about the nature of science. *Science Scope* 31 (2): 42–47.

Weiner, J. 1994. *The beak of the finch.* New York: Alfred Knopf.

What Is a Hypothesis?

Hypotheses are used widely in science. Put an X next to the statements that describe a hypothesis.

_____ **A** A tentative explanation

_____ **B** A statement that can be tested

_____ **C** An educated guess

_____ **D** An investigative question

_____ **E** A prediction about the outcome of an investigation

_____ **F** A question asked at the beginning of an investigation

_____ **G** A statement that may lead to a prediction

_____ **H** Included as a part of all scientific investigations

_____ **I** Used to prove whether something is true

_____ **J** Eventually becomes a theory, then a law

_____ **K** May guide an investigation

_____ **L** Used to decide what data to pay attention to and seek

_____ **M** Developed from imagination and creativity

_____ **N** Must be in the form of "if…then…"

Describe what a hypothesis is in science. Include your own definition of the word *hypothesis* and explain how you learned what it is.

_____

_____

_____

_____

What Is a Hypothesis?

Teacher Notes

What is a
Hypothesis ?

Purpose

The purpose of this assessment probe is to elicit students' ideas about hypotheses. The probe is designed to find out if students understand what a hypothesis is, when it is used, and how it is developed.

Related Concepts

hypothesis, nature of science, scientific inquiry, scientific method

Explanation

The best choices are A, B, G, K, L, and M. However, other possible answers open up discussions to contrast with the provided definition. A hypothesis is a tentative explanation that can be tested and is based on observation and/or scientific knowledge such as that that has been gained from doing background re-

search. Hypotheses are used to investigate a scientific question. Hypotheses can be tested through experimentation or further observation, but contrary to how some students are taught to use the "scientific method," hypotheses are not proved true or correct. Students will often state their conclusions as "My hypothesis is correct because my data prove...," thereby equating positive results with proof (McLaughlin 2006, p. 61). In essence, experimentation as well as other means of scientific investigation never *prove* a hypothesis—the hypothesis gains credibility from the evidence obtained from data that support it. Data either support or negate a hypothesis but never prove something to be 100% true or correct.

Hypotheses are often confused with questions. A hypothesis is not framed as a question but rather provides a tentative explanation in

response to the scientific question that leads the investigation. Sometimes the word *hypothesis* is oversimplified by being defined as "an educated guess." This terminology fails to convey the explanatory or predictive nature of scientific hypotheses and omits what is most important about hypotheses: their purpose. Hypotheses are developed to explain observations, such as notable patterns in nature; predict the outcome of an experiment based on observations or prior scientific knowledge; and guide the investigator in seeking and paying attention to the right data. Calling a hypothesis a "guess" undermines the explanation that underscores a hypothesis.

Predictions and hypotheses are not the same. A hypothesis, which is a tentative explanation, can lead to a prediction. Predictions forecast the outcome of an experiment but do not include an explanation. Predictions often use if-then statements, just as hypotheses do, but this does not make a prediction a hypothesis. For example, a prediction might take the form of, "If I do [X], then [Y] will happen." The prediction describes the outcome but it does not provide an explanation of why that outcome might result or describe any relationship between variables.

Sometimes the words *hypothesis, theory,* and *law* are inaccurately portrayed in science textbooks as a hierarchy of scientific knowledge, with the hypothesis being the first step on the way to becoming a theory and then a law. These concepts do not form a sequence for the development of scientific knowledge because each represents a different type of knowledge.

Not every investigation requires a hypothesis. Some types of investigations do not lend themselves to hypothesis testing through experimentation. A good deal of science is observational and descriptive—the study of biodiversity, for example, usually involves looking at a wide variety of specimens and maybe sketching and recording their unique characteristics. A biologist studying biodiversity might wonder, "What types of birds are found on island X?" The biologist would observe sightings of birds and perhaps sketch them and record their bird calls but would not be guided by a specific hypothesis. Many of the great discoveries in science did not begin with a hypothesis in mind. For example, Charles Darwin did not begin his observations of species in the Galapagos with a hypothesis in mind.

Contrary to the way hypotheses are often stated by students as an unimaginative response to a question posed at the beginning of an experiment, particularly those of the "cookbook" type, the generation of hypotheses by scientists is actually a creative and imaginative process, combined with the logic of scientific thought. "The process of formulating and testing hypotheses is one of the core activities of scientists. To be useful, a hypothesis should suggest what evidence would support it and what evidence would refute it. A hypothesis that cannot in principle be put to the test of evidence may be interesting, but it is not likely to be scientifically useful" (AAAS 1988, p. 5).

Curricular and Instructional Considerations

Elementary Students

In the elementary school grades, students typically engage in inquiry to begin to construct an understanding of the natural world. Their inquiries are initiated by a question. If students have a great deal of knowledge or have made prior observations, they might propose a hypothesis; in most cases, however, their knowledge and observations are too incomplete for them to hypothesize. If elementary school students are required to develop a hypothesis, it is often just a guess, which does little to contribute to an understanding of the purpose of a hypothesis. At this grade level, it is usually sufficient for students to focus on their questions, instead of hypotheses (Pine 1999).

Middle School Students

At the middle school level, students develop an understanding of what a hypothesis is and when one is used. The notion of a testable hypothesis through experimentation that involves variables is introduced and practiced at this grade level. However, there is a danger that students will think every investigation must include a hypothesis. Hypothesizing as a skill is important to develop at this grade level but it is also important to develop the understandings of what a hypothesis is and why and how it is developed.

High School Students

At this level, students have acquired more sci-entific knowledge and experiences and so are able to propose tentative explanations. They can formulate a testable hypothesis and demonstrate the logical connections between the scientific concepts guiding a hypothesis and the design of an experiment (NRC 1996).

Administering the Probe

This probe is best used as is at the middle school and high school levels, particularly if students have been previously exposed to the word *hypothesis* or its use. Remove any answer choices students might not be familiar with. For example, if they have not encountered if-then reasoning, eliminate this distracter. The probe can also be modified as a simpler version for students in grades 3–5 by leaving out some of the choices and simplifying the descriptions.

Related Ideas in *National Science Education Standards* (NRC 1996)

K–4 Understandings About Scientific Inquiry

- Scientific investigations involve asking and answering a question and comparing the answer with what scientists already know about the world.
- Scientists develop explanations using observations (evidence) and what they already know about the world (scientific knowledge).

5–8 Understandings About Scientific Inquiry

- Different kinds of questions suggest different kinds of investigations. Some investigations involve observing and describing objects, organisms, or events; some involve collecting specimens; some involve experiments; some involve seeking more information; some involve discovery of new objects and phenomena; and some involve making models.
- Current scientific knowledge and understanding guide scientific investigations. Different scientific domains employ different methods, core theories, and standards to advance scientific knowledge and understanding.

5–8 Science as a Human Endeavor

- Science is very much a human endeavor, and the work of science relies on basic human qualities such as reasoning, insight, energy, skill, and creativity.

9–12 Abilities Necessary to Do Scientific Inquiry

★ Identify questions and concepts that guide scientific investigations.

9–12 Understandings About Scientific Inquiry

- Scientists usually inquire about how physical, living, or designed systems function. Conceptual principles and knowledge guide scientific inquiries. Historical and current scientific knowledge influence the design and interpretation of investigations and the evaluation of proposed explanations made by other scientists.

Related Ideas in *Benchmarks for Science Literacy* (AAAS 1993)

K–2 Scientific Inquiry

- People can often learn about things around them by just observing those things carefully, but sometimes they can learn more by doing something to the things and noting what happens.

3–5 Scientific Inquiry

- Scientists' explanations about what happens in the world come partly from what they observe and partly from what they think. Sometimes scientists have different explanations for the same set of observations. That usually leads to their making more observations to resolve the differences.

6–8 Scientific Inquiry

★ Scientists differ greatly in what phenomena they study and how they go about their work. Although there is no fixed set of steps that all scientists follow, scientific investigations usually involve the collection of relevant evidence, the use of logical reasoning, and the application of imagination in devising hypotheses and explanations to make sense of the collected evidence.

★ Indicates a strong match between the ideas elicited by the probe and a national standard's learning goal.

6–8 Values and Attitudes

★ Even if they turn out not to be true, hypotheses are valuable if they lead to fruitful investigations.

9–12 Scientific Inquiry

★ Hypotheses are widely used in science for choosing what data to pay attention to and what additional data to seek and for guiding the interpretation of the data (both new and previously available).

Related Research

• Students generally have difficulty with explaining how science is conducted because they have had little contact with real scientists. Their familiarity with doing science, even at older ages, is "school science," which is often not how science is generally conducted in the scientific community (Driver et al. 1996).

• Despite over 10 years of reform efforts in science education, research still shows that students typically have inadequate conceptions of what science is and what scientists do (Schwartz 2007).

• Upper elementary school and middle school students may not understand experimentation as a method of testing ideas, but rather as a method of trying things out or producing a desired outcome (AAAS 1993).

• Middle school students tend to invoke personal experiences as evidence to justify their hypothesis. They seem to think of evidence as selected from what is already

known or from personal experience or secondhand sources, not as information produced through experiment (AAAS 1993).

Suggestions for Instruction and Assessment

• The "scientific method" is often the first topic students encounter when using textbooks and this can erroneously imply that there is a rigid set of steps that all scientists follow, including the development of a hypothesis. Often the scientific method described in textbooks applies to experimentation, which is only one of many ways scientists conduct their work. Embedding explicit instruction of the various ways to do science in the actual investigations students do throughout the year as well as in their studies of investigations done by scientists is a better approach to understanding how science is done than starting off the year with the scientific method in a way that is devoid of a context through which students can learn the content and process of science.

• Students often participate in science fairs that may follow a textbook scientific method of posing a question, developing a hypothesis, and so on, that incorrectly results in students "proving" their hypothesis. Make sure students understand that a hypothesis can be disproven, but it is never proven, which implies 100% certainty.

• Help students understand that science begins with a question. The structure of some school lab reports may lead students

---

★ Indicates a strong match between the ideas elicited by the probe and a national standard's learning goal.

to believe that all investigations begin with a hypothesis. While some investigations do begin with a hypothesis, in most cases, they begin with a question. Sometimes it is just a general question.

- A technique to help students maintain a consistent image of science as inquiry throughout the year by paying more careful attention to the words they use is to create a "caution words" poster or bulletin board (Schwartz 2007). Important words that have specific meanings in science but are often used inappropriately in the science classroom and through everyday language can be posted in the room as a reminder to pay careful attention to how students are using these words. For example, words like *hypothesis* and *scientific method* can be posted here. Words that are banned when referring to hypotheses include *prove, correct,* and *true.*

- Use caution when asking students to write lab reports that use the same format regardless of the type of investigation conducted. The format used in writing about an investigation may imply a rigid, fixed process or erroneously misrepresent aspects of science, such as that hypotheses are developed for every scientific investigation.

- Avoid using hypotheses with younger children when they result in guesses. It is better to start with a question and have students make a prediction about what they think will happen and why. As they acquire more conceptual understanding and experience a variety of observations, they will be better prepared to develop hypotheses that reflect the way science is done.

- Avoid using "educated guess" as a description for *hypothesis.* The common meaning of the word *guess* implies no prior knowledge, experience, or observations.

- Scaffold hypothesis writing for students by initially having them use words like *may* in their statements and then formalizing them with if-then statements. For example, students may start with the statement, "The growth of algae may be affected by temperature." The next step would be to extend this statement to include a testable relationship, such as, "If the temperature of the water increases, then the algae population will increase." Encourage students to propose a tentative explanation and then consider how they would go about testing the statement.

Related NSTA Science Store Publications and NSTA Journal Articles

American Association for the Advancement of Science (AAAS). 1993. *Benchmarks for science literacy.* New York: Oxford University Press.

Keeley, P. 2005. *Science curriculum topic study: Bridging the gap between standards and practice.* Thousand Oaks, CA: Corwin Press.

McLaughlin, J. 2006. A gentle reminder that a hypothesis is never proven correct, nor is a theory ever proven true. *Journal of College Science Teaching* 36 (1): 60–62.

National Research Council (NRC). 1996. *National science education standards.* Washington, DC: National Academy Press.

Schwartz, R. 2007. What's in a word? How word choice can develop (mis)conceptions about the nature of science. *Science Scope* 31 (2): 42–47.

VanDorn, K., M. Mavita, L. Montes, B. Ackerson, and M. Rockley. 2004. Hypothesis-based learning. *Science Scope* 27: 24–25.

Related Curriculum Topic Study Guides

(Keeley 2005)

"Inquiry Skills and Dispositions"

"Understandings About Scientific Inquiry"

References

American Association for the Advancement of Science (AAAS). 1988. *Science for all Americans.* New York: Oxford University Press.

American Association for the Advancement of Science (AAAS). 1993. *Benchmarks for science literacy.* New York: Oxford University Press.

Driver, R., J. Leach, R. Millar, and P. Scott. 1996. *Young people's images of science.* Buckingham, UK: Open University Press.

Keeley, P. 2005. *Science curriculum topic study: Bridging the gap between standards and practice.* Thousand Oaks, CA: Corwin Press.

McLaughlin, J. 2006. A gentle reminder that a hypothesis is never proven correct, nor is a theory ever proven true. *Journal of College Science Teaching* 36 (1): 60–62.

National Research Council (NRC). 1996. *National science education standards.* Washington, DC: National Academy Press.

Pine, J. 1999. To hypothesize or not to hypothesize. In *Foundations: A monograph for professionals in science, mathematics, and technology education. Vol. 2. Inquiry: Thoughts, views, and strategies for the K–5 classroom.* Arlington, VA: National Science Foundation.

Schwartz, R. 2007. What's in a word? How word choice can develop (mis)conceptions about the nature of science. *Science Scope* 31 (2): 42–47.

Life, Earth, and Space Science Assessment Probes

Life, Earth, and Space Science Assessment Probes

Concept Matrix

Related Science Concepts	Life Science					Earth and Space Science						
	Does It Have a Life Cycle?	Cells and Size	Sam's Puppy	Respiration	Rotting Apple	Earth's Mass	What Are Clouds Made Of?	Where Did the Water Come From?	Rainfall	Summer Talk	Me and My Shadow	Where Do Stars Go?
Cell Division			✓									
Cell Size		✓										
Cellular Respiration				✓								
Clouds							✓		✓			
Closed System						✓						
Condensation							✓	✓				
Conservation of Matter						✓						
Cycling of Matter						✓						
Decay (or Decomposition)					✓	✓						
Decomposers					✓							
Development	✓											
Earth-Sun System										✓	✓	
Evaporation								✓				
Gravity									✓			
Growth	✓		✓									
Life Cycle	✓											
Living vs. Nonliving	✓											
Microbes					✓							
Micrometer (or Micron)		✓										
Precipitation									✓			
Rain									✓			
Reproduction	✓											
Respiratory System				✓								
Seasons										✓		
Shadows											✓	
Stars												✓
Transformation of Matter			✓			✓						
Water Cycle							✓	✓	✓			
Water Vapor							✓	✓				

Does It Have a Life Cycle?

How do you decide if an organism goes through a life cycle? Put an X next to the organisms that have a life cycle.

_____ frog	_____ cow	_____ daisy
_____ butterfly	_____ mushroom	_____ chicken
_____ grasshopper	_____ grass	_____ maple tree
_____ fern	_____ earthworm	_____ human
_____ shark	_____ snail	_____ beetle
_____ bean plant	_____ mold	_____ crab
_____ snake	_____ spider	_____ moth

Explain your thinking. Describe the rule or reason you used to decide if an organism has a life cycle.

_____

_____

_____

_____

_____

_____

_____

_____

Does It Have a Life Cycle?

Teacher Notes

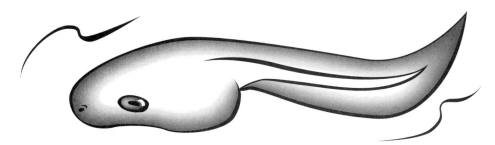

Purpose

The purpose of this assessment probe is to elicit students' ideas about life cycles. The probe can be used to determine whether students recognize that although life cycles vary in length and developmental stages, all multicellular organisms go through a life cycle.

Related Concepts

development, growth, life cycle, living vs. nonliving, reproduction

Explanation

All of the organisms on the list go through a life cycle. The entire lifespan of an organism, including the birth of a new generation of offspring, is called a *life cycle*. A life cycle typically includes fertilization and development of the embryo or embryo-like stage, birth or emer-gence, growth and development into an adult, reproduction, and death of the adult. It is cyclic because most adult organisms reproduce and give rise to new offspring, which keep the cycle going. At some stage in the life cycle of multicellular organisms, they stop reproducing and eventually die.

Stages of the life cycle vary among different types of organisms. For example, some organisms undergo changes during their early development in which the developing organism looks very different from the adult (e.g., butterfly, frog, beetle). Other organisms give rise to offspring with developing features that are similar to adult features (e.g., shark, human, maple tree, cow). Some life cycles are short (measured in days) and some are long (measured in years). Other differences include details of fertilization and zygote development.

Curricular and Instructional Considerations

Elementary Students

Elementary school students observe a variety of living organisms in the classroom to learn about their life cycles. Direct experiences include raising butterflies, frogs, and plants to study their life cycles. Representations are often used to sequence life cycles and to compare and contrast different types of cycles, such as complete and incomplete metamorphosis in insects. Studying the life cycle of an organism helps children understand the continuity of life.

Middle School Students

In middle school, students learn about fertilization (including pollination) as the beginning of an animal or plant's life cycle. Changes in the development of a plant or animal embryo are examined, including similarities between development of different species of plant or animal embryos. Details of human reproduction and development are introduced at the middle school level. At this level, students begin to link the idea of cell division to growth of an organism.

High School Students

In high school, biology students build on their basic K–8 understanding of sexual reproduction and development to focus on the haploid and diploid cellular details. They learn about complex life cycles of certain types of animals, fungi, and vascular and nonvascular plants, including alternation of generations (alternation of sexual and asexual reproduction) and sexual variations, such as parthenogenesis (development of an organism from an unfertilized egg), changing from male to female or vice versa, and hermaphrodism (having both male and female reproductive organs).

Administering the Probe

For younger students, you may choose to reduce the number of organisms on the list and/or include pictures of each. Remove any organisms on the list that students may be unfamiliar with. Consider adding additional items that students may have encountered in their local environment. This probe could be used with a card sort: Have students group items into those with life cycles and those without, and listen carefully to their reasoning. Extend the probe even further by asking students to describe the stages of the life cycle for each item they select. Listen carefully for indications that students recognize a cyclic process that includes being born, reproduction, and death, and do not just focus on the features of the developmental change each organism on the list goes through.

Related Ideas in *National Science Education Standards* (NRC 1996)

K–4 Life Cycles of Organisms

★ Plants and animals have life cycles that include being born, developing into adults, reproducing, and eventually dying. The

★ Indicates a strong match between the ideas elicited by the probe and a national standard's learning goal.

details of this life cycle are different for different organisms.

5–8 Reproduction and Heredity

- Reproduction is a characteristic of all living systems. Because no individual organism lives forever, reproduction is essential to the continuation of every species.
- In many species, including humans, females produce eggs and males produce sperm. Plants also reproduce sexually: The egg and sperm are produced in the flowers of the flowering plants. An egg and sperm unite to begin development of a new individual.

Related Ideas in *Benchmarks for Science Literacy* (AAAS 1993)

K–2 Diversity of Life

- Some animals and plants are alike in the way they look and the things they do, and others are very different from one another.

K–2 Human Development

- All animals have offspring; usually two parents are involved.
- A human baby grows inside its mother until its birth. Even after birth, a human baby is unable to care for itself, and its survival depends on the care it receives from adults.

K–2 Constancy and Change

- Things change in some ways and stay the same in some ways.

3–5 Human Development

- It takes about nine months for a human embryo to develop.
- Human beings live longer than most other animals, but all living things die.
- There is a usual sequence of stages in physical and mental development in human beings, although individuals differ in exactly when they reach each stage.

6–8 Human Development

- Fertilization occurs when sperm cells from a male's testes are deposited near an egg cell from the female ovary, and one of the sperm cells enters the egg cell.
- Following fertilization, cell division produces a small cluster of cells that then differentiate by appearance and function to form the basic tissues of an embryo. During the first three months of pregnancy, organs begin to form. During the second three months, all organs and body features develop. During the last three months, the organs and features mature enough to function well after birth. Patterns of human development are similar to those of other vertebrates.
- Various body changes occur as adults age.

9–12 Human Development

- The very long period of human development (compared with that of other species) is associated with the prominent role of the brain in human evolution.

Related Research

- In a study that investigated 10- to 14-year-old children's ideas about the continuity of life, most could correctly sequence pictures of seed germination, but 66% did not view the seed as alive and 19% did not understand the continuity of life from seed to seedling (Driver et al. 1994, p. 49).

- Some studies indicate that children fail to consider death as part of a life cycle (Driver et al. 1994).

- As students investigate the life cycles of organisms, teachers might observe that young children do not understand the continuity of life from, for example, seed to seedling or larvae to pupae to adult (NRC 1996, p. 128).

- Some K–8 students tend to equate life cycles only with the examples they observed in school, such as certain types of plant, butterfly, frog, or mealworm life cycles or organisms that are similar to those they studied. When students encounter organisms that are different from the ones they studied, they fail to recognize that all organisms have a life cycle (Authors' analysis of student work).

Suggestions for Instruction and Assessment

- When students make observations of a particular plant or animal's life cycle, be explicit in developing the generalization that all animals and plants go through a life cycle, even though details of their cycles differ.

- Use the term *continuity of life* along with *life cycle* so that the bigger idea of life continuing from generation to generation is emphasized. Teaching the stages in the life of different organisms is part of the idea of life cycles but can overshadow the more important idea of continuity if not explicitly addressed.

- Observe directly, if possible, or provide multiple visual examples of life cycles that differ in details. Have elementary school students identify the common pattern of birth, growth and development, reproduction, and death. Have middle and high school students identify fertilization, embryo development, birth or emergence, growth and development, reproduction, aging, and death.

- Avoid linear representations of life cycle stages that do not imply a cyclic process. Representations used should portray the continuity of life.

- So that students do not develop the misconception that life cycles begin after an egg hatches, a seedling emerges, or an animal gives birth, explicitly target the idea that seeds and eggs are alive and that there is a developing organism inside some animals.

- Emphasize the diversity among species in the details of their life cycles while pointing out the commonalities, not only between different animal species or different plant species, but between animal and plant species as well. Provide opportunities throughout students' K–12 experiences to examine a variety of life cycles, including organisms they may not be as familiar with.

Related NSTA Science Store Publications and Journal Articles

Ansberry, K., and E. Morgan. 2007. Loco beans. In *More picture-perfect science lessons,* 65–73. Arlington, VA: NSTA Press.

Cavallo, A. 2005. Cycling through plants. *Science and Children* (Apr./May): 22–27.

Driver, R., A. Squires, P. Rushworth, and V. Wood-Robinson. 1994. *Making sense of secondary science: Research into children's ideas.* London and New York: RoutledgeFalmer.

Keeley, P. 2005. *Science curriculum topic study: Bridging the gap between standards and practice.* Thousand Oaks, CA: Corwin Press.

Pliske, C. 2000. Natural cycles: Coming full circle. *Science and Children* (Mar.): 35–40.

Schussler, E., and J. Winslow. 2007. Drawing on students' knowledge. *Science and Children* (Jan.): 40–44.

Related Curriculum Topic Study Guide

(Keeley 2005)

"Reproduction, Growth, and Development (Life Cycles)"

References

American Association for the Advancement of Science (AAAS). 1993. *Benchmarks for science literacy.* New York: Oxford University Press.

Driver, R., A. Squires, P. Rushworth, and V. Wood-Robinson. 1994. *Making sense of secondary science: Research into children's ideas.* London and New York: RoutledgeFalmer.

Keeley, P. 2005. *Science curriculum topic study: Bridging the gap between standards and practice.* Thousand Oaks, CA: Corwin Press.

National Research Council (NRC). 1996. *National science education standards.* Washington, DC: National Academy Press.

Cells and Size

Which things are larger than cells? Put an X next to the things that are generally larger than a typical animal or plant cell.

_____ thickness of a leaf   _____ grain of salt

_____ atom   _____ protein molecule

_____ width of a hair   _____ DNA

_____ piece of sawdust   _____ eye of an ant

_____ water molecule   _____ tiny seed

_____ bread crumb   _____ larva of a tiny fruit fly   _____ speck of pepper

_____ bacteria   _____ period at end of sentence   _____ dust mite

_____ chromosome   _____ frog embryo   _____ virus

_____ point of a pin   _____ flea egg

Explain your thinking. How did you decide if something is larger than a typical plant or animal cell?

_____

_____

_____

_____

_____

_____

_____

_____

Cells and Size

Teacher Notes

Purpose

The purpose of this assessment probe is to elicit students' ideas about the size of cells. The probe can be used to determine whether students recognize how small a cell is relative to other things.

Related Concepts

cell size, micrometer (or micron)

Explanation

Although some of the choices depend on the size of a small object, the best choices are: thickness of a leaf, grain of salt, eye of an ant, width of a hair, piece of sawdust, tiny seed, bread crumb, larva of a tiny fruit fly, speck of pepper, period at end of a sentence, dust mite, frog embryo, point of a pin, and flea egg. Several of the items in this list are living things or parts of plants and animals and thus are made up of a collection of cells (the cell layers that make up the width of a leaf, sawdust, eye of an ant, tiny seed, larva of a tiny fruit fly, speck of pepper, dust mite, and microscopic frog embryo), which generally makes them larger than a single "typical" animal or plant cell. Chromosomes are organelles found within plant and animal cells, which make them smaller than a cell. Likewise, proteins, DNA, and water are molecules found within a cell and within cell structures, which reasons that they are also smaller than a cell. Bacteria are much smaller than animal and plant cells and viruses are much smaller than bacteria. The atom is the smallest particle of matter on the list. Trying to figure out the number of atoms in a cell is almost like trying to figure out the number of stars in the sky.

Generally, any very small object or particle of matter that can be seen with a hand-held magnifying lens or the human eye (which can detect sizes up to about 0.1 mm) is larger than a cell. However, an object or organism does not have to be visible by the unaided eye to be larger than a cell. To get a quantitative sense of scale, cells are typically measured in micrometers (μm; also called *microns*). There are 1,000 μm in 1 mm. Most types of plant and animal cells generally range between 10 and 100 μm (some cells, like eggs, nerve cells, and muscle cells, are much larger than "average" cells). The point of a pin is about 1,500 μm. A grain of table salt is about 300 μm. The width of a human hair is about 200 μm and a flea's egg is about 500 μm. Dust mites, bizarre looking multicellular animals, are still larger than typical cells even though they are typically photographed using electron microscopes. Several of these dust mites live at the base of your eyelashes and feed on secretions and dead skin cell debris! Dust mites range in size from 250 to 400 μm.

Typically, a microscope with magnification greater than 10× is needed to see most cells. Magnifications of 100× and more are needed to see things smaller than typical cells. While cells of different tissues vary in size, they are still much smaller than many of the things on the list. For example, a typical animal cheek cell is 60 μm, a red blood cell is about 8 μm, and a small leaf's cell is about 30 μm. Bacteria are single-celled organisms that are much smaller than a plant or animal cell. E. coli, a common bacterium, measures 2 μm.

The typical common cold virus measures 20 nm (nanometer). There are 1,000 nm in 1 μm. Even smaller is a water molecule. It measures about 0.2 nm!

Curricular and Instructional Considerations

Elementary Students

Students in the early elementary school grades use magnifying lenses to observe parts of living things that are too small to see clearly with their naked eye. Upper elementary students are just beginning to learn about cells and use simple microscopes to observe them. However, students' conceptions of a cell's very small size are limited by their ability to grasp very small magnitudes of scale.

Middle School Students

Students' fine motor skills help them become more adept in using compound microscopes at the middle school level to view a variety of cells and small parts of organisms and objects. They can interpret what they see under a microscope, can determine the magnification of their view, and begin to connect the size of cells to numbers that are much smaller than a millimeter. However, small scales are still difficult for them to comprehend.

High School Students

At the high school level, students transition from the whole cell to looking at structures within the cell. They learn about the molecules that make up a cell. Their understandings en-

compass smaller scales, including a growing awareness of nanoscale and nanoscience. They use more sophisticated microscopes and microscopic techniques that allow them to see bacterial cells.

Administering the Probe

This probe can be used once students understand that all organisms are made up of cells. Remove items on the list that may be unfamiliar to students.

Related Ideas in *National Science Education Standards* (NRC 1996)

· ·

K–4 Abilities Necessary to Do Scientific Inquiry

*   Employ simple equipment and tools (magnifiers and simple microscopes) to gather data and extend the senses.

5–8 Abilities Necessary to Do Scientific Inquiry

*   Use appropriate tools (microscopes) and techniques to gather, analyze, and interpret data.

5–8 Structure and Function in Living Systems

*   All organisms are composed of cells, the fundamental unit of life.
*   Groups of specialized cells cooperate to form a tissue, such as a muscle. Different tissues are grouped together to form larger functional units, called *organs*.

9–12 The Cell

*   Cells have particular structures that underlie their functions. Inside the cell is a concentrated mixture of thousands of different molecules that form a variety of specialized structures.
*   Most of the cells in a human contain two copies of each of 22 different chromosomes.
*   Each DNA molecule in a cell forms a single chromosome.

Related Ideas in Benchmarks for Science Literacy (AAAS 1993)

· ·

K–2 The Cell

*   Magnifiers help people see things that they would otherwise not be able to see.

K–2 Scale

*   Things in nature and things people make have very different sizes, weights, ages, and speeds.

3–5 The Cell

*   ★ Microscopes make it possible to see that living things are made mostly of cells. Some organisms are made of a collection of similar cells that benefit from cooperating.
*   Some living things consist of a single cell.

6–8 Cells

*   ★ All living things are composed of cells, from just one to many millions, whose details usually are visible only through a microscope. Different body tissues and

---

★ Indicates a strong match between the ideas elicited by the probe and a national standard's learning goal.

organs are made up of different kinds of cells.

9–12 Cells

- Within every cell are specialized parts.
- The work of the cell is carried out by the many different types of molecules it assembles, mostly proteins.

Related Research

- Studies have shown that students have difficulties with orders of magnitude. In a study of 16-year-old Israeli students (Dreyfus and Jungwirth 1988, 1989), students thought that molecules of protein were bigger than the size of a cell. Over a third of students' responses showed "inadequate" ideas about cells (Driver et al. 1994).
- Research conducted by Arnold (1983) indicated that students have difficulty differentiating between the concepts of cell and molecule. Students identified any materials encountered in biology class (carbohydrates, proteins, and water) as being made up of smaller units called *cells*. Arnold coined the term *molecell* to describe the notion of organic molecules being considered as cells.
- The range of numbers people can grasp increases with age (AAAS 1993, p. 276).

Suggestions for Instruction and Assessment

- Be aware that describing cells as being "very small" is a relative term to students. Small compared to what? When teaching the concept of smallness of cells, make comparisons to things that are smaller and larger than a cell.
- Provide students with opportunities to examine and compare very small things that their unaided eyes can detect, such as a grain of salt or width of a hair, to things their unaided eyes cannot detect, such as individual cells on a prepared microscope slide. Using a microscope, have students compare the two differently sized things, noting the difference in relative size under the same magnification.
- Begin by having students in the early elementary grades use 3×–10× magnification hand lenses to magnify things. Encourage them to wonder what they might see with more powerful lenses (AAAS 1993).
- By the upper elementary school grades, the magnification that students use should increase to 30×–100× magnification, using more powerful handheld viewers, dissection scopes, or simple microscopes. Students' observations should include microscopic one-celled organisms, plant and animal cells, and small animals, such as brine shrimp. As they observe different types of cells, they should be encouraged to think about whether those cells could be seen without a microscope.
- By middle school, students should have developed "magnification sense" and can extend their observations of cells using a microscope to photographs of cells, including bacteria, taken under much greater magnifications than their school microscopes can provide.

- Middle school students are learning about the fundamental unit of matter (the atom) and molecules made up of atoms as well as the basic unit of life (the cell). When taught separately, there is the potential for misconceptions to develop related to the sizes of atoms, molecules, and cells. Explicitly address the size of a cell in comparison to atoms and molecules because some students think they are similar in size and fail to recognize that cells are made up of atoms and molecules. The Powers of Ten website, at *www.powersof10.com,* provides a source of representations to help students distinguish between the magnitudes of scale in observing cells versus atoms and molecules.

- At the high school level, help students make explicit comparisons of the size of protein and DNA molecules to the size of a cell, recognizing that these molecules fit within cells. Combining this probe with "Is It Made of Cells?" and "Is It Made of Molecules?" from Volume 1 of this series (Keeley, Eberle, and Farrin 2005) may help reveal whether high school students hold the concept of molecell, as is indicated by research studies.

Related NSTA Science Store Publications and Journal Articles

American Association for the Advancement of Science (AAAS). 1993. *Benchmarks for science literacy.* New York: Oxford University Press.

Driver, R., A. Squires, P. Rushworth, and V. Wood-Robinson. 1994. *Making sense of secondary sci-*

Robinson. 1994. *Making sense of secondary science: Research into children's ideas.* London and New York: RoutledgeFalmer.

Jones, G., M. Falvo, A. Taylor, and P. Broadwell. 2007. *Nanoscale science: Activities for grades 6–12.* Arlington, VA: NSTA Press.

Keeley, P. 2005. *Science curriculum topic study: Bridging the gap between standards and practice.* Thousand Oaks, CA: Corwin Press.

National Research Council (NRC). 1996. *National science education standards.* Washington, DC: National Academy Press.

Related Curriculum Topic Study Guide

(Keeley 2005)

"Cells"

References

American Association for the Advancement of Science (AAAS). 1993. *Benchmarks for science literacy.* New York: Oxford University Press.

Arnold, B. 1983. Beware the molecell! Aberdeen College of Education. *Biology Newsletter* 42: 2–6.

Dreyfus, A., and E. Jungwirth. 1988. The cell concept of 10th graders: Curricular expectations and reality. *International Journal of Science Education* 10 (2): 221–229.

Dreyfus, A., and E. Jungwirth. 1989. The pupil and the living cell: A taxonomy of dysfunctional ideas about an abstract idea. *Journal of Biological Education* 23 (1): 49–53.

Driver, R., A. Squires, P. Rushworth, and V. Wood-Robinson. 1994. *Making sense of secondary sci-*

ence: *Research into children's ideas.* London and New York: RoutledgeFalmer.

Keeley, P. 2005. *Science curriculum topic study: Bridging the gap between standards and practice.* Thousand Oaks, CA: Corwin Press.

Keeley, P., F. Eberle, and L. Farrin. 2005. *Uncovering student ideas in science: 25 formative assessment probes.* Vol. 1. Arlington, VA: NSTA Press.

National Research Council (NRC). 1996. *National science education standards.* Washington, DC: National Academy Press.

Sam's Puppy

Sam brought home a tiny puppy. Her puppy grew. Four weeks later, her puppy had grown to twice its original size. Which answer below best explains why Sam's puppy got bigger?

A The number of cells in the puppy's body increased.

B The puppy's body absorbed the food it ate.

C The puppy's body cells grew larger as it got older.

D Parts of the puppy's body stretched out more.

Describe your thinking. Provide an explanation for your answer.

_____

_____

_____

_____

_____

_____

_____

_____

_____

_____

_____

Sam's Puppy

Teacher Notes

Purpose

The purpose of this assessment probe is to elicit students' ideas about growth. The probe can be used to determine whether students recognize that growth occurs as a result of cell division, which increases the number of body cells.

Related Concepts

cell division, growth, transformation of matter

Explanation

The best response is A: The number of cells in the puppy's body increased. Body cell reproduction involves producing new daughter cells for growth of tissues as well as repair and replacement of old cells. *Growth* is the term for the overall increase in an organism's size. It is a complex process, but to describe it in simple terms, growth primarily involves cell enlarge-

ment as new molecules are added to the cell's mass and subsequent cell division. Food and nutrients taken in by the puppy are broken down at a molecular level, transformed within cells, and become the building blocks for new living material, including new cells produced through cell division. Proteins are a main constituent of living tissues in animals and one of the most important raw materials for growth. During growth, the molecules that result from the breaking down of food, such as amino acids from proteins or sugars from carbohydrates, are synthesized into new molecules within cells, adding more molecules and thus more mass to the structures that make up an organism's body.

Most living body cells eventually divide into two cells through a process called *mitosis*. During mitosis, a body cell enlarges, duplicates its genetic material, and divides into two daughter

cells. Sometimes the daughter cells are smaller than the original cell and do not become as large as the original cell until new molecules are synthesized within the new cell. However, most body cells generally remain the same size and do not continuously grow larger as an organism develops (there are some exceptions, such as muscle cells). Growth determines not only the size of the puppy but also its shape and form. As long as the puppy grows at the same rate along all its dimensions, its bodily proportions remain generally the same. One part generally does not "stretch out" more than another.

Curricular and Instructional Considerations

Elementary Students

Elementary school students observe a variety of living organisms in the classroom to learn about their life cycles. Growth and development are necessary parts of understanding life cycles. At this grade level, growth is understood at a macroscopic level and connected to the needs of organisms, such as food being a requirement for growth. Students can observe and measure an organism's growth, but a cellular and molecular explanation is not expected until middle school. However, the probe is useful in determining children's preconceived ideas about growth.

Middle School Students

In middle school, students learn how food provides the building material for all organisms and that it can be transformed and made part of a growing organism's body. Students

develop basic understandings of cell structure and function. They learn how cells divide to make more cells. The topic of cell division is often taught as a mechanism and memorized as a series of steps and not explicitly linked to the idea of adding new molecules that result in increased body mass and growth. In their study of reproduction, students learn how an egg and sperm unite and that subsequent cell division and differentiation begin the development of the organism. They examine how an organism grows and develops until it reaches adulthood. Because the human organism is of great interest at this age level, middle school curricula often focus on the growth and development of humans and looking for similar patterns in other vertebrate organisms.

High School Students

In high school biology class, students build on basic cell division ideas that were developed in middle school. They learn about details related to cell differentiation and division and how these processes are regulated. They examine how cell division occurs in different types of tissues and the effect of aging and other factors on cell division and growth. A focus on molecular biology helps students understand how food is chemically broken down into the chemical constituents cells need to synthesize other molecules, which contribute to an organism's mass as it grows.

Administering the Probe

This probe can be used once students understand that all organisms are made up of cells.

The distracters are intentionally kept simple in order to elicit a range of ideas from elementary grades through high school. At the middle school and high school level, the puppy in the probe can be replaced with a human baby, and students can be asked to compare growth from the time the baby was brought home after birth to the same child as a toddler one year later. Photographs showing a "baby" organism and the same organism after it has considerably grown can be used as props. This probe can be combined with the probe "Whale and Shrew" in Volume 2 of this series (Keeley, Eberle, and Tugel 2007) to further explore students' ideas regarding cell size.

Related Ideas in *National Science Education Standards* (NRC 1996)

. . . .

K–4 Life Cycles of Organisms

★ Plants and animals have life cycles that include being born, developing into adults, reproducing, and eventually dying. The details of this life cycle are different for different organisms.

5–8 Structure and Function in Living Systems

★ Cells carry on the many functions needed to sustain life. They grow and divide, thereby producing more cells. This requires that organisms take in nutrients, which cells use to provide energy for the work that they do and to make the materials that they or their organism needs.

9–12 The Cell

- Most cell functions involve chemical reactions. Food molecules taken into cells react to provide the chemical constituents needed to synthesize other molecules.

- Cell functions are regulated. Regulation occurs both through changes in activity of the functions performed by proteins and through selective expression of individual genes. This regulation allows cells to respond to their environment and to control and coordinate cell growth and division.

Related Ideas in *Benchmarks for Science Literacy* (AAAS 1993)

. . . .

3–5 Basic Functions

- People obtain energy and materials from food for body repair and growth.

3–5 Flow of Matter and Energy

- Some source of energy is needed for all organisms to stay alive and grow.

6–8 Cells

★ Cells continually divide to make more cells for growth and repair.

- Food provides the fuel and building material for all organisms.

6–8 Basic Functions

- For the body to use food for energy and building materials, the food must first be digested into molecules that are absorbed and transported to cells.

★ Indicates a strong match between the ideas elicited by the probe and a national standard's learning goal.

9–12 Cells

- Complex interactions among the different kinds of molecules in the cell cause distinct cycles of activities such as growth and division.

Related Research

- Students of all ages think food is a requirement for growth rather than a source of matter for growth. They have little knowledge about food being transformed and made part of a growing organism's body (AAAS 1993).

- Several studies have shown students' difficulties in assimilating taught concepts of cell division, cell enlargement, and cell differentiation. One study showed that only 69% of respondents who were asked what accounts for growth realized that growth was occurring when a cell divides in two (Driver et al. 1994).

- Children understand at an early age that eating or absorbing food is necessary for growth. However, they do not recognize that these materials are the material for growth and that they are transformed and taken into the body, thus making it bigger (Driver et al. 1994).

- A study by Russell and Watt (1990) found that children think animals grow or stretch to accommodate the food they eat.

Suggestions for Instruction and Assessment

- At the middle school level and above, explicitly connect the idea of growth to cell division. Also, explicitly connect the idea of food to the material that makes up the matter of a growing organism. Often these concepts—cell division, food, and growth—are taught separately without explicit links made between them.

- Avoid teaching mitosis as a sequence of mechanistic steps in somatic cell division. Help students link the process of mitosis to growth (as well as repair) of an organism.

- For older students, particularly students who think growth occurs because cells keep getting larger, use activities that demonstrate why cells cannot carry out their functions if they grow to a very large size.

- Recognize that *grow,* in the common meaning of the word, means "gets bigger." Help students understand the biological meaning of *grow.*

- Do not separate teaching about growth from teaching about development. Growth and development are essential components of understanding life cycles. When students are ready to utilize cell concepts, build on elementary students' macroscopic observations of growth and development by having them investigate and explain growth and development at the cellular level.

- To get a sense of the amount of matter resulting from cell divisions, have students mathematically determine exponential increases to see how rapidly numbers of cells can build up as they undergo cell division. Starting with a single cell, have students calculate or graph how many new cells can

result in one day from cell divisions every 30 minutes.

- To go from macroscopic ideas about growth to microscopic ideas, have students compare sizes and increase in mass of organs in adults versus organs in children. Probe students to think about and explain how additional mass is added as a result of adding additional molecules. Since the original cells cannot contain all these additional molecules, new cells are added that account for the increase in mass, and thus size.

Related NSTA Science Store Publications and Journal Articles

American Association for the Advancement of Science (AAAS). 1993. *Benchmarks for science literacy.* New York: Oxford University Press.

Driver, R., A. Squires, P. Rushworth, and V. Wood-Robinson. 1994. *Making sense of secondary science: Research into children's ideas.* London and New York: RoutledgeFalmer.

Keeley, P. 2005. *Science curriculum topic study: Bridging the gap between standards and practice.* Thousand Oaks, CA: Corwin Press.

National Research Council (NRC). 1996. *National science education standards.* Washington, DC: National Academy Press.

Related Curriculum Topic Study Guides

(Keeley 2005)

"Cells"

"Reproduction, Growth, and Development (Life Cycles)"

References

American Association for the Advancement of Science (AAAS). 1993. *Benchmarks for science literacy.* New York: Oxford University Press.

Driver, R., A. Squires, P. Rushworth, and V. Wood-Robinson. 1994. *Making sense of secondary science: Research into children's ideas.* London and New York: RoutledgeFalmer.

Keeley, P. 2005. *Science curriculum topic study: Bridging the gap between standards and practice.* Thousand Oaks, CA: Corwin Press.

Keeley, P., F. Eberle, and J. Tugel. 2007. *Uncovering student ideas in science: 25 more formative assessment probes.* Vol. 2. Arlington, VA: NSTA Press.

National Research Council (NRC). 1996. *National science education standards.* Washington, DC: National Academy Press.

Russell, T., and D. Watt. 1990. *Growth: Primary SPACE Project, Research Report.* Liverpool, UK: Liverpool University Press.

Respiration

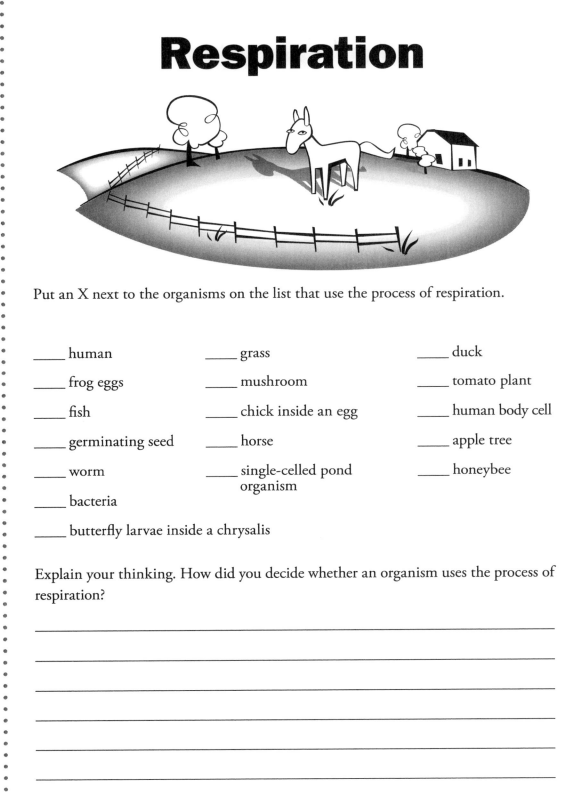

Put an X next to the organisms on the list that use the process of respiration.

_____ human	_____ grass	_____ duck
_____ frog eggs	_____ mushroom	_____ tomato plant
_____ fish	_____ chick inside an egg	_____ human body cell
_____ germinating seed	_____ horse	_____ apple tree
_____ worm	_____ single-celled pond organism	_____ honeybee
_____ bacteria		
_____ butterfly larvae inside a chrysalis		

Explain your thinking. How did you decide whether an organism uses the process of respiration?

_____

_____

_____

_____

_____

_____

Respiration

Teacher Notes

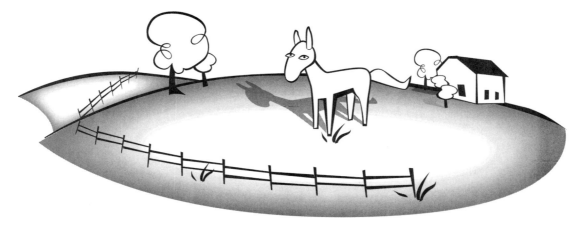

Purpose

The purpose of this assessment probe is to elicit students' ideas about respiration. The probe is designed to find out whether students recognize respiration as a process that all living things use in order to obtain energy or whether they have a restricted macroscopic meaning of respiration.

Related Concepts

cellular respiration, respiratory system

Explanation

Everything on the list uses the process of respiration. Respiration is an essential life process carried out by all living organisms—from single-celled to multicelled—to provide the energy that organisms need to function. Aerobic respiration happens at two levels. At the organism level, it generally involves taking in the air that contains the oxygen needed by cells and eliminating carbon dioxide from the body. At the cellular level, the oxygen is used to break down molecules of food in order to release the energy needed by cells to function. Carbon dioxide is released by the cell as a waste product.

Most people, including students, commonly understand that animals with some form of a respiratory system breathe in oxygen from the air through their respiratory systems and breathe out carbon dioxide. They usually equate the gas exchange during aerobic respiration with breathing rather than a cellular process. All animals respire, but they are not the only organisms to do so. Because every living organism is composed of at least one cell, and all cells need energy to function, then every

organism must carry out some form of cellular respiration regardless of whether it has a respiratory system that includes organs such as lungs or gills.

While different types of organisms may perform respiration in different ways, all organisms use respiration to release energy through the breakdown of molecules within a cell. Aerobic respiration involves an interchange of gases between an organism and its environment. Sometimes this interchange involves multicelled structures (e.g., organs) in an organism that take in oxygen and make it available to the cells. For example, plants take in oxygen through their leaves and animals take in oxygen through their lungs or gills where it is sent to and used within their cells to break down sugars (food) to release energy. Single-celled organisms can absorb oxygen into a cell directly from the environment. Respiration can also occur in the absence of oxygen. This type of anaerobic respiration occurs with some types of bacteria and fungi as well as in the muscle cells of animals when there is a lack of oxygen.

Organisms in an immature stage of development, such as the butterfly larvae in a chrysalis, frog eggs, and a chick developing inside an egg, also respire by taking oxygen, making it available to their cells, and releasing energy from food molecules. They are all living things that need energy to develop. Under the right conditions of temperature and moisture, seeds respire by taking in oxygen, although they can be dormant for long periods of time before germination. Plants utilize oxygen in the process of respiration. They also take in carbon dioxide and release oxygen in the process of photosynthesis. However, these two processes are not opposites nor are they mutually exclusive. Respiration in plants also happens during photosynthesis.

Curricular and Instructional Considerations

Elementary Students

At the elementary school level, students distinguish between living and nonliving things and learn that most living things need air. Respiration at this level is usually equated with breathing and focuses on familiar structures of animals and plants that take in oxygen, such as lungs, gills, and leaves. As students investigate single-celled organisms, they learn that these simple organisms also need air.

Middle School Students

In middle school, students continue to learn about various structures that take in oxygen and make it available to cells, including the structures of insects and aquatic organisms. At this level, they begin to connect the taking in of oxygen to the needs of cells, developing a basic understanding of cellular respiration without going into the details of cell structure and biochemical processes. Students connect the need of cells for oxygen to their growing understanding of oxidation as a process that releases energy from food within cells. At this stage, students should begin to develop the generalization that all organisms respire, since energy is needed by all living things.

High School Students

In high school biology class, students build on their basic middle school understanding of cellular respiration to examine the process at the cellular and molecular level, including the eukaryotic structures involved, such as mitochondria as well as prokaryotic cellular respiration. They learn about and distinguish between the processes of aerobic respiration and anaerobic respiration. However, at this level, as students learn about the process of photosynthesis in more detail, some may believe that only animals respire and that photosynthesis is the opposite process in plants.

Administering the Probe

Eliminate items from the list that students may not be familiar with, or explain each one, showing pictures if they are unsure as to what the organism is. For elementary school students who are not expected to know about cellular respiration, consider adapting this probe by using familiar language, such as, "Does it use air?" and reducing the number of choices. For high school students, consider adding other nonanimal choices, such as "algae" and "virus," and replace "human body cell" with specific types of cells.

Related Ideas in *National Science Education Standards* (NRC 1996)

· ·

K–4 The Characteristics of Organisms

★ Organisms have basic needs. For example, animals need air, water, and food; plants require air, water, nutrients, and light.

5–8 Structure and Function in Living Systems

★ Cells carry out the many functions needed to sustain life. They grow and divide, thereby producing more cells. This requires that they take in nutrients, which they use to provide energy for the work that cells do and to make the materials that a cell or organism needs.

9–12 The Cell

· Cells have particular structures that underlie their functions. Inside the cell is a concentrated mixture of thousands of different molecules that form a variety of different structures that carry out such cell functions as energy production.

9–12 Matter, Energy, and Organization in Living Systems

· Living systems require a continuous input of energy to maintain their chemical and physical organizations.

· The chemical bonds of food molecules contain energy. Energy is released when the bonds of food molecules are broken down and new compounds with lower energy bonds are formed.

Related Ideas in *Benchmarks for Science Literacy* (AAAS 1993)

· ·

K–2 The Cell

★ Most living things need water, food, and air.

★ Indicates a strong match between the ideas elicited by the probe and a national standard's learning goal.

3–5 The Cell

★ Some living things consist of a single cell. Like familiar organisms, they need food, water, and air; a way to dispose of waste; and an environment they can live in.

3–5 The Human Organism

• By breathing, people take in the oxygen they need to live.

6–8 The Cell

• Within cells, many of the basic functions of organisms, such as extracting energy from food and getting rid of waste, are carried out. The way in which cells function is similar in all living organisms.

6–8 Flow of Matter and Energy

★ Animals get energy from oxidizing their food, releasing some of its energy as heat.

6–8 The Human Organism

★ To burn food for the release of energy stored in it, oxygen must be supplied to cells and carbon dioxide must be removed. Lungs take in oxygen for the combustion of food and they eliminate the carbon dioxide produced.

9–12 The Cell

• Within every cell are specialized parts for the transport of materials, energy transfer, protein building, waste disposal, information feedback, and even movement.

Related Research

• Although students have ideas about gas exchange and usually equate it with breathing, few students at any age have a complete understanding of respiration (Driver et al. 1994).

• In a study by Haslam and Treagust (1987), most students thought respiration and breathing were synonymous.

• Studies have found that although young children recognize air as being necessary for life, they have limited understanding of what happens to air once it is inhaled. Many think only organisms with lungs use air. Few students, all the way through high school, connect food with the use of oxygen (Driver et al. 1994).

• Some students think oxygen is the gas needed by animals and carbon dioxide is the gas needed by plants. Some students think photosynthesis is the plant version of energy release or that respiration only happens in animals (Driver et al. 1994).

• Some students think plants only use oxygen (respire) in the dark (Driver et al. 1994).

Suggestions for Instruction and Assessment

• When teaching students the idea that organisms need air (or oxygen), explicitly address the variety of ways that organisms take in air (or oxygen) so that students do not equate exchange of gases only with animals that have lungs. Furthermore, when students learn about various structures that allow multicellular organisms to take

---

★ Indicates a strong match between the ideas elicited by the probe and a national standard's learning goal.

in oxygen, be sure to address how single-celled organisms also take in oxygen.

- Photosynthesis is not the opposite of respiration. The intake and release of gases in the two processes is opposite, but the processes themselves are not opposites. Teaching this idea of opposites may perpetuate the notion that animals use oxygen (and thus respire) and plants do not and that gas flow in plants always happens in opposite directions.

- For students in middle school, explicitly connect the idea of cells taking in oxygen to the need to release energy from food.

- Be careful when using a burning analogy to describe how energy is released from food. Oxidation reactions involving food are not combustion reactions.

- The idea that seeds, frog's eggs, and the chrysalis of a butterfly do not respire or use air may be connected to students' ideas about living or nonliving. In order to accept the idea that these stages in an organism's life respire or need air, they need to accept that they are living stages. It may be useful to combine this probe with the probe from Volume 1 of this series, "Is It Living?" (Keeley, Eberle, and Farrin 2005).

Related NSTA Science Store Publications and Journal Articles

American Association for the Advancement of Science (AAAS). 1993. *Benchmarks for science literacy.* New York: Oxford University Press.

Driver, R., A. Squires, P. Rushworth, and V. Wood-Robinson. 1994. *Making sense of secondary sci-*

ence: Research into children's ideas. London and New York: RoutledgeFalmer.

Keeley, P. 2005. *Science curriculum topic study: Bridging the gap between standards and practice.* Thousand Oaks, CA: Corwin Press.

Littlejohn, P. 2007. Building leaves and an understanding of photosynthesis. *Science Scope* 8 (30): 22–25.

National Research Council (NRC). 1996. *National science education standards.* Washington, DC: National Academy Press.

Related Curriculum Topic Study Guide

(Keeley 2005)
"Photosynthesis and Respiration"

References

American Association for the Advancement of Science (AAAS). 1993. *Benchmarks for science literacy.* New York: Oxford University Press.

Driver, R., A. Squires, P. Rushworth, and V. Wood-Robinson. 1994. *Making sense of secondary science: Research into children's ideas.* London and New York: RoutledgeFalmer.

Haslam, F., and D. Treagust. 1987. Diagnosing secondary students' misconceptions of photosynthesis and respiration in plants using a two-tier multiple choice instrument. *Journal of Biological Education* 21 (3): 203–211.

Keeley, P. 2005. *Science curriculum topic study: Bridging the gap between standards and practice.* Thousand Oaks, CA: Corwin Press.

Keeley, P., F. Eberle, and L. Farrin. 2005. *Uncovering student ideas in science: 25 formative assessment*

probes. Vol. 1. Arlington, VA: NSTA Press.

National Research Council (NRC). 1996. *National science education standards.* Washington, DC: National Academy Press.

Rotting Apple

Four friends argued about why an apple on the ground eventually rots away and disappears. This is what they said:

Anna: "I think it is just something that happens over time."

Selma: "I think small organisms use it for energy and building material."

Felicia: "I think the atoms and molecules in the apple just break apart."

Logan: "I think wind and water soften it, and it dissolves into the soil."

Eli: "I think water and air rot it, then small animals come and eat the rest."

Jack: "I think it gets old and breaks apart into pieces too small to see."

Which student do you most agree with? _____

Describe your thinking. Provide an explanation for your answer.

_____

_____

_____

_____

_____

_____

_____

_____

_____

_____

_____

Rotting Apple

Teacher Notes

Purpose

The purpose of this assessment probe is to elicit students' ideas about decay and decomposers. The probe can be used to determine whether students recognize the need for a biological agent to break down once-living material as it uses it for energy.

Related Concepts

decay, decomposers, decomposition, microbes

Explanation

The best response is Selma's: I think small organisms use it for energy and building material. Rotting, or decay, is a natural recycling process in which the material of once-living things is broken down and released into the environment to be reassembled and used again by other organisms or incorporated into the physical environment. Natural decay does not happen without a biological agent. Dead organisms or parts of once-living organisms are broken down by biological agents, called *decomposers*. The material of the apple that is no longer part of the living tree is a source of food for decomposers. This food provides the energy and building material they need to live, grow, and reproduce. Decomposers include worms, insects, fungi, and bacteria. Most decay is the result of action by microorganisms (bacteria and some fungi). Decomposers are found in large numbers almost everywhere in an ecosystem, including in air, land, and water. They are prevalent in the upper layers of most soils. Fungal spores and bacteria also make up particulate matter in the air. The factors that support decay are the factors that support microbial growth, such as warmth and moisture.

Curricular and Instructional Considerations

. . . .

Elementary Students

During the elementary grades, children build an understanding of biological concepts through directly observing phenomena, such as the rotting apple. Students identify factors that promote decay and observe the changes that happen to a decomposing object over time. They develop a basic understanding of decomposers and the decay process, beginning with macroscopic organisms they can readily observe, such as worms, beetles, mushrooms, and molds.

Middle School Students

In the middle school grades, students become familiar with the beneficial role of bacteria in ecosystems. They expand their understanding of decomposers and decay to include microorganisms and recognize their essential role in the decomposition process as matter recyclers. They are introduced to ideas about nutrition, matter, and energy flow, and they identify the relationships between organisms in a food web, including decomposers.

High School Students

In high school, students approach decomposition from a molecular view, including the assortment of complex biological processes involved in breaking down once-living material. They recognize the role of decomposers in cycling atoms and molecules throughout the living and nonliving components of Earth's biosphere. They connect the natural process of biodegradation to human engineered systems that solve problems of buildup of dead material and metabolic wastes.

Administering the Probe

The answer to this probe deliberately uses the idea that small organisms use the apple for energy and building material rather than directly mentioning food, but it does not include the idea of breaking it down into simpler substances since students can choose that response to match their notion of rotting, yet not understand the decay process. When *food* is used in the correct response, students are apt to choose it without considering the idea of decay. To students, an apple is food and other organisms eat apples. Because the idea of a source of energy is introduced in upper elementary school, we choose to use *energy* in this answer rather than *food*. However, if you are using this probe with younger students, grades K–2, you may wish to substitute *food* for *energy and building material* and remove Felicia's response. This probe can also be used with a prop. Show a rotted apple (preferably one that has been on the ground) and a fresh one. Other material can be substituted for the apple (e.g., pumpkins, rotted logs, dead leaves). Avoid bringing in dead and decaying animals or decayed logs with exposed fungi that can release spores.

Related Ideas in *National Science Education Standards* (NRC 1996)

K–4 The Characteristics of Organisms

• Organisms have basic needs.

K–4 Organisms and Their Environments

• All organisms cause changes in the environments where they live. Some of these changes are detrimental to the organism or to other organisms, whereas others are beneficial.

5–8 Populations and Ecosystems

★ Decomposers, primarily bacteria and fungi, are consumers that use waste materials and dead organisms for food.

9–12 The Interdependence of Organisms

• Energy flows through ecosystems in one direction, from photosynthetic organisms to herbivores to carnivores and decomposers.

9–12 Matter, Energy, and Organization in Living Systems

• As matter and energy flow through different levels of organization of living systems—cells, organs, organisms, communities—and between living systems and the physical environment, chemical elements are recombined in different ways.

Related Ideas in *Benchmarks for Science Literacy* (AAAS 1993)

K–2 Flow of Matter and Energy

• Many materials can be recycled and used again, sometimes in different forms.

K–2 Constancy and Change

• Things change in some ways and stay the same in some ways.

3–5 Interdependence of Life

★ Insects and various other organisms depend on dead plant and animal material for food.

• Most microorganisms do not cause disease and many are beneficial.

3–5 Flow of Matter and Energy

• Some source of energy is needed for all organisms to stay alive and grow.

• Over the whole Earth, organisms are growing, dying, and decaying, and new organisms are being produced by the old ones.

6–8 Interdependence of Life

★ Two types of organisms may interact with one another in several ways: They may be in a producer-consumer, predator-prey, or parasite-host relationship. Or, one organism may scavenge or decompose another.

6–8 Flow of Matter and Energy

• Food provides molecules that serve as fuel and building material for all organisms.

★ Indicates a strong match between the ideas elicited by the probe and a national standard's learning goal.

9–12 Flow of Matter and Energy

- The amount of life any environment can support is limited by the available energy, water, oxygen, and minerals, as well as by the ability of ecosystems to recycle the residue of dead organic materials.

- The chemical elements that make up the molecules of living things pass through food webs and are combined and recombined in different ways.

Related Research

- Several research studies have identified children's common notions about decay. In response to research questions related to the "disappearance" of dead animals or fruits on the surface of the soil, young children thought dead things just disappear (Driver et al. 1994).

- Some middle school students see decay as a gradual, inevitable consequence of time without need of decomposing agents (AAAS 1993).

- Younger students tend to think that insects break up material once it has started to rot of its own accord (Driver et al. 1994).

- In a study of 15- to 16-year-olds, 65% used words like *bacteria, fungi,* and *decomposers,* but were not sure about their roles. Although older students tend to offer more factors to explain decay, there was little evidence that they had an understanding of how physical factors relate to the action of microbes (Driver et al. 1994, p. 65).

- Generally, students are unaware of the role that microorganisms play in ecosystems—especially their role as agents of change, such as decomposers and recyclers (Driver et al. 1994).

- When fifth and sixth graders were asked what makes a dead thing disappear, some of their comments included, "When it's been dead a long time and gets real old, it breaks up and disappears," "When the rain and wind come, the dead plant spreads out into the dirt," "When we die they put us in a coffin and bury us, and while we're in the coffin we dissolve" (Hogan 1994).

- Some students say things "decay" or "rot away" without realizing that microorganisms cause the decay process as they use dead material for food (Hogan 1994).

Suggestions for Instruction and Assessment

- In activities involving observations of changes that happen during decomposition, explicitly link the notion of a living agent that causes the changes to the changes students observe.

- Emphasize the notion that all living things need food, including microorganisms. Food can be the sugars produced by plants and the fresh forms of plants, animals, and certain fungi we are familiar with. However, food for other organisms can be what we find offensive—rotting material and waste products. Develop the commonality that all of these sources of food serve to provide living organisms—from single-celled organisms to large plants and animals—the energy and building material for growth and repair.

- Decay can be observed carefully by collecting dead leaves in a woodland habitat and separating them into different layers. The top layer usually consists of the whole, dead leaves. The bottom layer usually consists of the broken-down humus. In between are leaves in various stages of decay. Explicitly link observations of the changes in the leaves to the organisms responsible for the changes.

- The primary agents of decomposition are microorganisms, and because they are not visible to the naked eye, students tend to overlook them as agents of decay. Students tend to associate microorganisms with disease and less with beneficial processes like decay of dead plants and animals. Explicitly develop an appreciation of the role of microorganisms as decomposers and recyclers.

- With young children, be aware that focusing on and investigating the role of one organism, such as worms, in breaking down material may lead students to a narrow view of what kinds of organisms are considered to be decomposers.

- Explicitly ask students what makes things rot and draw a picture or create storyboards of rotting, explaining what happens in each stage. Look for evidence of physical processes, such as the action of heat and water versus biological processes.

- Elicit students' conception of what food is. Ask them if a rotten apple or a decaying animal is food and probe further for their ideas.

- Encourage students to develop operational

definitions of *decay* and *decomposition* before introducing the scientific terminology.

- Use safe practices when bringing in decaying plant material for students to observe. Some students have allergic reactions to mold spores. It is best to place rotting material in a closed container for student observations.

- Make connections to human-designed systems that take advantage of decomposition, such as composting.

Related NSTA Science Store Publications and Journal Articles

Ashbrook, P. 2004. Teaching through trade books: Meet the decomposers. *Science and Children* (Summer): 14–16.

Driver, R., A. Squires, P. Rushworth, and V. Wood-Robinson. 1994. *Making sense of secondary science: Research into children's ideas.* London and New York: RoutledgeFalmer.

Keeley, P. 2005. *Science curriculum topic study: Bridging the gap between standards and practice.* Thousand Oaks, CA: Corwin Press.

Trautman, N., and Environmental Inquiry Team. 2003. *Decay and renewal.* Arlington, VA: NSTA Press.

Related Curriculum Topic Study Guide
(Keeley 2005)

"Decomposers and Decay"

References

American Association for the Advancement of Science (AAAS). 1993. *Benchmarks for science literacy.* New York: Oxford University Press.

Driver, R., A. Squires, P. Rushworth, and V. Wood-Robinson. 1994. *Making sense of secondary science: Research into children's ideas.* London and New York: RoutledgeFalmer.

Hogan, K. 1994. *Eco-inquiry: A guide to ecological learning experiences for the upper elementary/ middle grades.* Millbrook, NY: Institute of Ecosystem Studies.

Keeley, P. 2005. *Science curriculum topic study: Bridging the gap between standards and practice.* Thousand Oaks, CA: Corwin Press.

National Research Council (NRC). 1996. *National science education standards.* Washington, DC: National Academy Press.

Earth's Mass

In autumn, dead leaves fall off trees. Every day, animals eliminate waste. All plants and animals eventually die. As a result, what happens to the mass of the Earth? Circle the best response.

A The mass of the Earth steadily decreases.

B The mass of the Earth steadily increases.

C The mass of the Earth stays about the same.

Explain your thinking. Describe the rule or reasoning you used to select your answer.

_____

_____

_____

_____

_____

_____

_____

_____

_____

_____

_____

_____

_____

_____

Earth's Mass

Teacher Notes

Purpose

The purpose of this assessment probe is to elicit students' ideas about the cycling of matter. The probe can be used to determine whether students recognize that once-living matter breaks down and cycles through ecosystems without subtracting or adding mass to the Earth.

Related Concepts

closed system, conservation of matter, cycling of matter, decay, transformation of matter

Explanation

The best response is C: The mass of the Earth stays about the same. Although some mass is added to the Earth by meteorites and micrometeorites, it is so minuscule that for the purpose of this probe, it can be neglected. Likewise, some mass is lost when hydrogen at-

oms at the edge of our atmosphere escape into space, nuclear reactors convert matter to energy, or rockets and other materials are launched into space. But, this too is negligible as far as the total mass of the Earth. The reason why the Earth's mass stays about the same, even though tons of dead and decomposing material are produced each minute, is explained by the conservation of matter principle. No matter what physically or chemically happens to materials in a closed system, the amount of matter, and thus the mass, stays the same. No new matter is added or taken away. Earth is primarily considered a closed system in regard to matter (not energy), even though there are some small amounts of material entering or leaving from space. This means that the amount of material on the Earth stays pretty much the same, even though its form, compo-

sition, and location can change.

Earth does not receive new inputs of elements, such as carbon, hydrogen, oxygen, nitrogen, silicon, calcium, and so on, that make up living and nonliving things. What we see is what we get. The amount of matter that made up our original Earth is still here today. Matter is continuously transformed through biological, physical, and geological processes. Carbon and other nutrients constantly cycle between living and nonliving things. For example, plants use carbon dioxide and water molecules to make sugars. These sugars are transformed into plant material and, say, eaten by a rabbit. Some material is converted back into the inorganic carbon dioxide and water of the atmosphere when the rabbit respires. Furthermore, when organisms, like the rabbit, die or parts of organisms, like leaves, fall to the ground, they are broken down by decomposers such as bacteria, worms, and insects and "recomposed."

As decomposers use food from dead material, gases are released back into the atmosphere through respiration, and molecules are incorporated into the decomposer's cell and body structures. Some of the material is further broken down and released as waste into the soil, water, or air. The materials within the Earth may be further transformed by geological processes into rocks and minerals. Matter in the soil, air, and water may be taken in again by other living organisms and transformed into a new material.

The most important thing to keep in mind is that living and nonliving matter never disappears (it may disappear from sight but it does not cease to exist) or is added as additional matter to the Earth's total mass. When organisms die or waste materials are produced, the number of atoms remains the same even though the material seems to disappear or build up. All matter can be accounted for through various transformations.

Curricular and Instructional Considerations

Elementary Students

During the elementary school grades, children build an understanding of recycling that forms the foundational idea for later understandings about the cycling of living or once-living matter. The idea that materials can be reused in different forms begins with objects and extends to once-living things in the upper elementary grades. In those grades, students develop a basic understanding of decomposers and the decay process, beginning with macroscopic organisms they can observe, such as worms, beetles, mushrooms, and molds. They begin to notice that substances can change form and move from place to place but they never appear out of nowhere and never just disappear (AAAS 1993, p. 119).

Middle School Students

In the middle school grades, students become familiar with ecosystems, including the beneficial role of bacteria in ecosystems. They expand their understanding of decomposers and decay to include microorganisms and recognize the essential role of microorganisms in the decom-

position process as matter recyclers. They are introduced to ideas about nutrition and matter and energy flow and identify the relationships between organisms in a food web, including decomposers. At this level, they begin to trace matter as it moves through ecosystems and should connect it to the notion of atoms. The idea of systems is made more explicit and they can now link the concept of closed systems and cycling of matter to the conservation of matter principle.

High School Students

At this level, students should have a sufficient grasp of atoms and molecules so as to link the conservation of matter with the flow of energy in living systems (AAAS 1993, p. 121). In high school, students approach decomposition from a molecular view, including the assortment of complex biological processes involved in breaking down once-living material. They recognize the role of decomposers in cycling atoms and molecules throughout the living and nonliving components of Earth's biosphere while conserving matter. Students connect the natural process of biodegradation to human-engineered systems that solve problems of buildup of dead material and metabolic waste.

Administering the Probe

This probe can be combined with "Rotting Apples" and the conservation of matter probes in Volume 1 of *Uncovering Student Ideas in Science* (Keeley, Eberle, and Farrin 2005). The word *weight* can be substituted for *mass* without compromising the ideas elicited by

the probe. Substitute *weight* if the concept of mass is not well developed with younger students, because it can interfere with students' interpretation of the probe. Research indicates that some students confuse the word *mass* with the phonetically similar word *massive* and thus equate the probe with size rather than amount of matter.

Related Ideas in *National Science Education Standards* (NRC 1996)

K–4 Organisms and Their Environments

- All organisms cause changes in the environment where they live. Some of these changes are detrimental to the organism or other organisms, whereas others are beneficial.

5–8 Populations and Ecosystems

- Decomposers, primarily bacteria and fungi, are consumers that use waste materials and dead organisms for food.

9–12 The Interdependence of Organisms

★ The atoms and molecules on the Earth cycle among the living and nonliving components of the biosphere.

9–12 Matter, Energy, and Organization in Living Systems

- As matter and energy flow through different levels of organization of living systems—cells, organs, organisms, communities—and between living systems and the

---

★ Indicates a strong match between the ideas elicited by the probe and a national standard's learning goal.

physical environment, chemical elements are recombined in different ways.

9–12 Geochemical Cycles

★ The Earth is a system containing essentially a fixed amount of each stable atom or element. Each element can exist in several different chemical reservoirs. Each element on Earth moves among reservoirs in the solid earth, oceans, atmosphere, and organisms as part of geochemical cycles.

9–12 Natural Resources

• Natural systems have the capacity to reuse waste, but that capacity is limited.

Related Ideas in *Benchmarks for Science Literacy* (AAAS 1993)

. . . .

K–2 Flow of Matter and Energy

• Many materials can be recycled and used again, sometimes in different forms.

K–2 Constancy and Change

• Things change in some ways and stay the same in some ways.

3–5 Interdependence of Life

• Insects and various other organisms depend on dead plant and animal material for food.

3–5 Flow of Matter and Energy

★ Over the whole Earth, organisms are growing, dying, and decaying, and new organ-

isms are being produced by the old ones.

6–8 Structure of Matter

★ No matter how substances within a closed system interact with one another, or how they combine or break apart, the total weight of the system remains the same. The idea of atoms explains the conservation of matter: If the number of atoms stays the same no matter how they are rearranged, then their total mass stays the same.

6–8 Flow of Matter and Energy

★ Over a long time, matter is transferred from one organism to another repeatedly and between organisms and their physical environment. As in all material systems, the total amount of matter remains constant, even though its form and location change.

6–8 Systems

• A system can include processes as well as things.
• A system is usually connected to other systems, both internally and externally. Thus a system may be thought of as containing subsystems and as being a subsystem of a larger system.

9–12 Flow of Matter and Energy

• At times, environmental conditions are such that plants and marine organisms grow faster than decomposers can recycle them back to the environment. Layers of energy-rich organic material have been gradually turned into great coal beds and

---

★ Indicates a strong match between the ideas elicited by the probe and a national standard's learning goal.

oil pools by the pressure of the overlying earth. By burning these fossil fuels, people are passing most of the stored energy back into the environment as heat and releasing large amounts of carbon dioxide.

★ The chemical elements that make up the molecules of living things pass through food webs and are combined and recombined in different ways.

Related Research

- Several research studies have identified children's common notions about decay. In response to research questions related to the "disappearance" of dead animals or fruits on the surface of the soil, young children thought dead things just disappear and did not allow for ideas about conservation of matter after death. Most of these children thought of decomposition as the total or partial disappearance of matter (Driver et al. 1994).

- Some middle school students think dead organisms simply rot away. They do not realize that matter from dead organisms is converted into other materials in the environment (AAAS 1993).

- A study by Leach et al. (1992) revealed that 70% of 11- to 13-year-olds lacked an understanding of conservation of matter to explain what happens after organisms die, even after the topic is taught. Furthermore, they found that up to age 16, few students had a view of matter that involved conservation in a variety of contexts, including decay (Driver et al. 1994).

- A study by Smith and Anderson (1986) found that middle school students seem to know that some kind of cyclical process takes place in ecosystems, but they tend to see only chains of events and pay little attention to the matter processes involved. Students tend to think the processes involve creating and destroying matter rather than chemically transforming it from one substance to another (AAAS 1993).

- Some middle school students recognize recycling of material through soil minerals but fail to incorporate water, oxygen, and carbon dioxide into matter cycles. Even after specially designed instruction, students cling to the misinterpretation that materials are recycled primarily through soil in the form of minerals (AAAS 1993).

- Generally, students are unaware of the role that microorganisms play in ecosystems, especially microorganisms' role as agents of change, such as decomposers and recyclers (Driver et al. 1994).

- Field-test results of this probe with middle school students revealed that students chose the correct response, but had incorrect reasons for why the mass of the Earth would stay the same. The most common incorrect justifications described a cycle of birth and death whereby every organism that dies is replaced by a new one born, thus no new matter is created or destroyed. Another common explanation lacks a cycling concept: When Earth was formed, there was a certain amount of matter, and when organisms die, the matter remains but is locked

---

★ Indicates a strong match between the ideas elicited by the probe and a national standard's learning goal.

up in dead organisms. As Earth produces more organisms, more matter is used until eventually no more new organisms can be produced and all the matter will be in dead organisms and waste products.

Suggestions for Instruction and Assessment

- Instruction that traces matter through the ecosystem as a basic pattern of thinking may help correct difficulties students have in understating conservation of matter in ecosystems. Students should be encouraged to consider where substances come from and where they go (AAAS 1993).

- Explicitly link the notion of the breakdown of molecules and the reassembly of atoms to food webs and cycles of matter in ecosystems.

- The notion of reusable building blocks common to plants and animals is difficult to understand if students do not first have an understanding of atoms and molecules. Once they grasp a particulate model of matter, have students use their understanding of atoms and molecules to follow matter through ecosystems (AAAS 1993).

- It is important for students to understand the concept of a closed system. Help students understand why Earth is considered a closed system and to use systems reasoning to explain why Earth does not gain or lose matter.

- Use examples of oil and coal beds to show ways in which material has been changed by the natural environment, other than recy-

cling by decomposers. Help students see how burning large reservoirs of fossil fuels does not decrease the mass of Earth, but releases the matter primarily in the form of carbon dioxide back into Earth's atmosphere.

- Use decomposition chambers, such as those developed by Bottle Biology, *www.bottlebiology.org*, to design investigations that show that matter in a closed system is conserved during the decay process.

Related NSTA Science Store Publications and Journal Articles

American Association for the Advancement of Science (AAAS). 1993. *Benchmarks for science literacy.* New York: Oxford University Press.

American Association for the Advancement of Science (AAAS). 2001. *Atlas of science literacy.* Vol. 1, "flow of matter in ecosystems," 76–77. Washington, DC: AAAS.

Driver, R., A. Squires, P. Rushworth, and V. Wood-Robinson. 1994. *Making sense of secondary science: Research into children's ideas.* London and New York: RoutledgeFalmer.

Keeley, P. 2005. *Science curriculum topic study: Bridging the gap between standards and practice.* Thousand Oaks, CA: Corwin Press.

Trautman, N. and Environmental Inquiry Team. 2003. *Decay and renewal.* Arlington, VA: NSTA Press.

Related Curriculum Topic Study Guides

(Keeley 2005)

"Conservation of Matter"
"Cycling of Matter in Ecosystems"
"Decomposers and Decay"

References

American Association for the Advancement of Science (AAAS). 1993. *Benchmarks for science literacy.* New York: Oxford University Press.

Driver, R., A. Squires, P. Rushworth, and V. Wood-Robinson. 1994. *Making sense of secondary science: Research into children's ideas.* London and New York: RoutledgeFalmer.

Keeley, P. 2005. *Science curriculum topic study: Bridging the gap between standards and practice.* Thousand Oaks, CA: Corwin Press.

Keeley, P., F. Eberle, and L. Farrin. 2005. *Uncovering student ideas in science: 25 formative assessment probes.* Vol. 1. Arlington, VA: NSTA Press.

Leach, J., R. Driver, P. Scott, and C. Wood-Robinson. 1992. *Progression in conceptual understanding of ecological concepts by pupils age 5–16.* Leeds, England: Center for Studies in Science and Mathematics Education, University of Leeds.

National Research Council (NRC). 1996. *National science education standards.* Washington, DC: National Academy Press.

Smith, E., and C. Anderson. 1986. Alternative student conceptions of matter cycling in ecosystems. Paper presented at the annual meeting of the National Association of Research in Science Teaching (NARST), San Francisco.

What Are Clouds Made Of?

Six friends were looking at large, white, puffy clouds in the sky. They wondered what the clouds were made of. This is what they said:

Trista: "I think clouds are made of large drops of water."

Lee: "I think clouds are made of soft, cottonlike material."

Manny: "I think clouds are made of smoke that rises from the land."

Rosie: "I think clouds are made of evaporated water in the form of a gas."

Glenda: "I think clouds are made of tiny drops of water or tiny ice crystals."

Leticia: "I think clouds are made of a spongy material that holds water in it."

Which friend do you most agree with? _____

Describe your thinking. Explain the reason for your answer.

_____

_____

_____

_____

_____

_____

What Are Clouds Made Of?

Teacher Notes

Purpose

The purpose of this assessment probe is to elicit students' ideas about an everyday object in the sky: clouds. The probe is designed to determine whether students recognize that clouds are made up of tiny droplets of water or tiny ice crystals.

Related Concepts

clouds, condensation, water cycle, water vapor

Explanation

The best response is Glenda's: I think clouds are made of tiny drops of water or tiny ice crystals. Clouds come in a variety of forms and shades. Not all are puffy and white. Some can be white and wispy while others are dark and appear to cover the entire sky. Regardless of shape, size, and shade, all clouds are formed when water vapor in the air cools, condenses, and becomes tiny drops of liquid water or tiny ice crystals. When water vapor condenses in the sky it becomes visible as a cloud. Cumulus clouds—the puffy, cottonlike clouds that are relatively low in the sky—are made up of billions and billions of tiny water drops. Other clouds, like the feathery cirrus clouds that are high in the sky, are made up of tiny ice crystals. The water drops and ice crystals are too small to see individually but they are just the right size to scatter the light that strikes them, making the clouds appear white. Rain clouds appear gray and contain bigger water drops. Rain eventually falls when the drops get too big to be held by the rising air that formed the cloud in the first place.

Curricular and Instructional Considerations

. . . .

Elementary Students

In the early elementary grades the emphasis should be on observing and describing clouds as well as other forms of visible water in the air, such as fog and mist. Their study of matter includes observing how water can go back and forth between different states. They begin to link ideas about states of matter to the water cycle and use their conceptual understanding of ice, liquid water, and water vapor to describe water in the air they can see (clouds and fog) and water in a form they cannot see (the air that surrounds them).

Middle School Students

Middle school students expand on their elementary experiences in observing and describing clouds to more conceptual ideas about the composition and formation of clouds. By middle school, students should know that liquid water or ice crystals in the air are visible and water vapor is not visible. The concept of evaporation is better understood by students at this level than the concept of condensation. These processes are combined with a growing understanding of the behavior of particles in the solid, liquid, and gas state. In addition, their knowledge of the properties of water helps complete their understanding of the water cycle.

High School Students

At the high school level, students should know what clouds are made of and how they are formed. This probe can be useful in diagnosing whether students have an understanding of this aspect of the movement of matter (water) in the Earth system. They expand their knowledge about Earth's atmosphere to an understanding of the Earth as a dynamic system. The water cycle is one of the aspects of that system. Students' growing knowledge of chemistry helps them to appreciate the mechanism of condensation at the particle level. They examine cloud formation at a more complex level and the global implications, including the effect of such aerosols as salt crystals, sand or soil particles, dust, smoke, or volcanic ash on forming cloud condensation nuclei that provide water vapor with a surface to condense upon.

Administering the Probe

All students have experienced seeing clouds in the sky (although in some geographic areas, clouds are more common in the sky on a daily basis). If possible, take students outside to view clouds or show a picture of a cumulus cloud to prompt their thinking before responding to the probe. Further probing can include a picture of a white cloud and a dark cloud; ask students if these clouds are made of the same material and ask them to describe what each cloud is made of. This probe can be used with other probes in this book, such as "Rainfall" (p. 171) and "Where Did the Water Come From?" (p. 163), or combined with "Wet Jeans" from Volume 1 of this series (Keeley, Eberle, and Farrin 2005) to create a cluster of water cycle–related probes.

Related Ideas in *National Science Education Standards* (NRC 1996)

K–4 Properties of Objects and Materials

- Materials can exist in different states, as a solid, liquid, or gas. Some common materials, such as water, can be changed from one state to another by heating or cooling.

K–4 Objects in the Sky

- The Sun, Moon, stars, clouds, birds, and airplanes all have properties, locations, and movements that can be observed and described.

5–8 Structure of the Earth System

- Water evaporates from the Earth's surface, rises and cools as it moves to higher elevations, condenses as rain or snow, and falls to the surface where it collects in lakes, oceans, soil, and in rocks underground.
- ★ Clouds, formed by the condensation of water vapor, affect weather and climate.

Related Ideas in *Benchmarks for Science Literacy* (AAAS 1993)

K–2 The Earth

- Water can be a liquid or a solid and can go back and forth from one form to another.

3–5 The Earth

- ★ When liquid water disappears, it turns into

a gas (vapor) in the air and can reappear as a liquid when cooled or as a solid if cooled below the freezing point of water. Clouds and fog are made of tiny droplets [or frozen crystals] of water. (Note: The brackets indicate language added to the original benchmark. This revised benchmark appears in AAAS 2007, p. 21.)

6–8 The Earth

- Water evaporates from the surface of the Earth, rises and cools, condenses into rain or snow, and falls again to the surface.

Related Research

- A study by Bar (1986) examined a sample of students ages 5–15 for their conceptions about aspects of the water cycle. He found certain ideas are more prevalent with certain ages. When students ages 5–7 were asked what causes rain, there was little evidence of a relationship between clouds and rain. Several students described clouds as being made of smoke or cotton wool. Of those that did see a link between rain and clouds, clouds were often described as bags of water kept high in the sky; when they collide, they rip open and the water falls out. Students ages 6–8 often described clouds as collecting water from the oceans and then moving to places in the sky above land. At ages 6–9, several students described clouds as being made of water vapor from the Sun heating the sea or water vapor that comes from kettles. At ages 7–10, some students visualize a cloud

---

★ Indicates a strong match between the ideas elicited by the probe and a national standard's learning goal.

as a sponge that has drops of water in it. The common idea at ages 9 and 10 is that clouds are made of water evaporated from puddles. At ages 11–15, students begin to describe clouds as made of small drops of water and water vapor that gets cold (Driver et al. 1994).

- A study by Phillips (1991) found that students at the high school level had beliefs similar to younger students. For example, students believed that clouds are formed by boiling—vapors from a kettle or the Sun boiling the sea; clouds are mostly smoke; clouds are made of cotton or wool; or clouds are bags of water.

- Field-test results of this probe with fourth through eighth graders revealed the commonly held idea that clouds are made of evaporated water in the form of water vapor. Students describe water evaporating from bodies of water as part of the water cycle and eventually forming clouds, but they failed to link the idea that the water vapor condensed to form clouds.

Suggestions for Instruction and Assessment

- Provide younger students with opportunities to make long-term observations of clouds and encourage them to talk about what they think they are made of.

- Have students research different types of clouds. The descriptions of different clouds often include their composition. Be aware that it is more common for students to know that clouds are made up of tiny drop-

lets of water than it is for them to know that they can be made of tiny droplets of water or tiny ice crystals.

- With young students, beware of art-related activities that use cotton balls to simulate clouds. This can lead to the misconception that clouds are made of a cottonlike substance.

- To compare condensation in the air (fog or clouds) with condensation against a surface (such as the dew on the grass in the morning), observe what happens inside a large glass jar. Put a spoonful of water in a large jar filled with air. With the jar in a warm place, vigorously swirl and shake the jar, pour the water out and put the top of the jar on upside down. This will raise the humidity inside the jar. Make sure there is no (or minimal) water on the walls of the jar. Put ice cubes in the upside-down top of the jar. Students will see the condensation on the inside of the lid and around the top of the jar but the air inside the jar will be relatively clear. This is similar to the dew forming on the grass in the morning while the air is clear.

Allow the jar to warm up, and repeat the procedure, only this time light a match and drop it into the jar. As soon as the match goes out, put the lid on the jar and place ice cubes in the lid. As it cools, students will see the fog form at the top of the jar and slowly sink to the bottom. The jar will not be clear. Discuss what happened and why it is different from what happened in the first jar (this time the water vapor con-

densed on the suspended smoke particles). Relate this to the formation of fog near the ground and clouds higher up (Foster 1991).

- Connect the idea of fog formation to dew point. Have students explain why fog tends to form in the evening or is visible in the morning but generally not during the day.

- When teaching about the water cycle, be careful not to overemphasize the terms *evaporation, condensation,* and *precipitation* at the expense of understanding what is actually happening to the water during these processes both in terms of its physical form and its location. Many students believe that the water evaporates to form a cloud and is still in the form of water vapor and that rainfall is the result of condensation.

- Ask students to draw a sequence of pictures to show and explain how they think clouds form.

- Be aware of poor diagrams of the water cycle that often show water evaporating and rising to form a "white cloud" and then moving to a "dark cloud." While the picture is intended to show changes in the same cloud before it rains, to a student it looks like two different clouds made up of seemingly different material—one made of evaporated water and the other made of condensed water. Representations that look like this are pervasive on the web and teachers need to be aware of the misconception these representations can promote.

- Older students can research technological applications of increasing the possibility

of rain cloud seeding by shooting small particles up into the clouds to create more condensation nuclei.

Related NSTA Science Store Publications and NSTA Journal Articles

American Association for the Advancement of Science (AAAS). 1993. *Benchmarks for science literacy.* New York: Oxford University Press.

American Association for the Advancement of Science (AAAS). 2007. *Atlas of science literacy.* Vol. 2, "weather and climate," 20–21. Washington, DC: AAAS.

Crane. P. 2004. On observing the weather. *Science and Children* (May): 32–36.

Driver, R., A. Squires, P. Rushworth, and V. Wood-Robinson. 1994. *Making sense of secondary science: Research into children's ideas.* London and New York: RoutledgeFalmer.

Keeley, P. 2005. *Science curriculum topic study: Bridging the gap between standards and practice.* Thousand Oaks, CA: Corwin Press.

National Research Council (NRC). 1996. *National science education standards.* Washington, DC: National Academy Press.

Robertson, W. 2005. *Air, water, and weather: Stop Faking It! Finally Understanding Science So You Can Teach It.* Arlington, VA: NSTA Press.

Vowell, J., and M. Phillips. A drop through time. *Science and Children* 44 (9):30–34.

Williams, J. 1997. *The weather book: An easy-to-understand guide to the USA's weather.* 2nd ed. Arlington, VA: NSTA Press.

Related Curriculum Topic Study Guides

(Keeley 2005)

"Air and Atmosphere"

"Weather and Climate"

References

American Association for the Advancement of Science (AAAS). 1993. *Benchmarks for science literacy.* New York: Oxford University Press.

American Association for the Advancement of Science (AAAS). 2007. *Atlas of science literacy.* Vol. 2, "weather and climate," 20–21. Washington, DC: AAAS.

Bar, V. 1986. *The development of the conception of evaporation.* Jerusalem, Israel: The Amos de-Shalit Science Teaching Center in Israel, The Hebrew University of Jerusalem, Israel.

Driver, R., A. Squires, P. Rushworth, and V. Wood-Robinson. 1994. *Making sense of secondary science: Research into children's ideas.* London and New York: RoutledgeFalmer

Foster, G. 1991. Fakey fog. In *Water, stones, and fossil bones,* ed. K. Lind, 78–79. Arlington, VA: NSTA Press.

Keeley, P. 2005. *Science curriculum topic study: Bridging the gap between standards and practice.* Thousand Oaks, CA: Corwin Press.

Keeley, P., F. Eberle, and L. Farrin. 2005. *Uncovering student ideas in science: 25 formative assessment probes.* Vol. 1. Arlington, VA: NSTA Press.

National Research Council (NRC). 1996. *National science education standards.* Washington, DC: National Academy Press.

Phillips, W. C. 1991. Earth science misconceptions. *The Science Teacher* 58 (2): 21–23.

Where Did the Water Come From?

Latisha took a sealed, plastic container of ice cubes out of the freezer. The outside of the container was dry when she took it out of the freezer. She set the container on the counter. She did not open the container. Half an hour later she noticed the ice had melted inside the container. The container was full of water. A small puddle of water had formed on the kitchen countertop, around the outside of the container. Which best describes where the puddle of water came from?

A A gas in the air.

B Melted ice inside the container.

C Cold on the outside of the container.

D Condensation from water inside the container.

E Water that evaporated from inside the container.

F Cold changed hydrogen and oxygen atoms to water.

Describe your thinking about where the water came from. Provide an explanation for your answer.

_____

_____

_____

_____

_____

Where Did the Water Come From?

Teacher Notes

Purpose

The purpose of this assessment probe is to elicit students' ideas about condensation. The probe is designed to determine whether students recognize that condensation comes from the water vapor in the air.

Related Concepts

condensation, evaporation, water cycle, water vapor

Explanation

The best response is A: A gas in the air. The phenomenon described in the probe is the condensation that occurs on the outside of a cold object when the object comes in contact with warmer air that contains water vapor. Water vapor is an invisible gas found in the air around us. When the air molecules containing water vapor come in contact with a cold object, the water vapor in the air changes to liquid water on the cold object. It is this water that comes from a gas in the air called *water vapor,* that forms the puddle. Water condenses on the outside of the container and drips down to form a puddle. At a molecular level, when the molecules of water vapor come in contact with a cold object like the container of ice cubes, they slow down and move closer together. As the molecules move closer together, the attraction between molecules increases and they cluster together to form liquid water.

This phenomenon appears in several everyday examples, such as the condensation that occurs on your bathroom mirror after a hot shower; the wet dew on the morning and evening grass; the wet glass of lemonade poured on a hot, humid day; the wetness on the outside of a window of a cool, air-conditioned house; and the wetness on the inside of a car window during the cold winter when water vapor from exhaled breath condenses on the cold windows.

Curricular and Instructional Considerations

Elementary Students

In the early elementary school grades, the emphasis is on observing and describing observable phenomena, such as the condensation that occurs on the outside of a cold object. Students should have multiple opportunities in different contexts throughout the elementary grades to observe changes in state, such as solid to liquid, liquid to solid, liquid to gas, and gas to liquid. As students progress to the upper elementary school grades, they begin to develop explanations for these phenomena. An explanation of condensation is combined with an explanation of evaporation—that water leaves open containers and bodies of water and exists in the air around us in a gas form we cannot see called *water vapor*. The molecular explanation of evaporation and condensation can wait until middle school. More important than developing the vocabulary words *condensation* and *evaporation* is first developing a conceptual understanding of the processes linked to observable phenomena. Once students understand what is happening to the water, the vocabulary can be introduced and used with meaning.

Middle School Students

The water cycle is of profound importance for middle school students' understanding of Earth systems. However, before the idea of the cyclic nature of water is developed, teachers need to be sure students understand what happens to water during a change in state. By the end of eighth grade, students should be able to use their understanding of the motion and position of molecules to explain changes in state and properties of solids, liquids, and gases. Many middle school students use the terms *evaporation* and *condensation,* which are words introduced at the elementary school level, without completely understanding where the water goes after it evaporates and why it condenses. Furthermore, students at this age still have difficulty accepting that air is matter and that there is water in the air in a form we cannot see. In order to explain everyday phenomena such as the one described in this probe, several of these ideas need to be integrated at the middle school level.

High School Students

During high school, students develop more sophisticated ideas about the mechanism of condensation at a molecular level and the cycling of matter, such as water, through ecosystems. They learn how matter moves around Earth in simple and complex molecules in vapor, solid, and liquid form and that the movement of matter is driven by the internal and external energy of Earth. Knowledge of heat, change in state, evaporation, condensation, and the force of gravity helps students understand *why* the water cycle occurs. However, do not assume that students have a correct conception of processes such as evaporation and condensation. Before

more complex ideas are taught in high school Earth science, this probe is useful in determining whether students have progressed beyond their preconceptions about where water goes after it evaporates and how it can reappear as a liquid.

Administering the Probe

Response A intentionally does not use the term *water vapor,* because students may choose this answer without knowing that water vapor is one of several gaseous substances that make up air. This probe scenario can be demonstrated to students if there is enough humidity in the classroom. Be aware that some students who live in dry, desert areas may not have experienced this phenomenon to the extent that students in warm, humid areas have. Consider having students draw a picture to support their explanation. When using this probe with elementary school students who may be unfamiliar with atoms and elements, consider whiting out the last distracter. This probe can be combined with "What Are Clouds Made Of?" (p. 155) and "Rainfall" (p. 171) from this volume and with "Wet Jeans" from Volume 1 of this series (Keeley, Eberle, and Farrin 2005) to make up a cluster of water cycle–related probes.

Related Ideas in *National Science Education Standards* (NRC 1996)

. . . .

K–4 Changes in Earth and Sky

- Materials can exist in different states, as a solid, liquid, or gas. Some materials, such as water, can be changed from one state to another by heating or cooling.

5–8 Structure of the Earth System

★ Water, which covers the majority of Earth's surface, circulates through the crust, oceans, and atmosphere in what is known as the *water cycle.* Water evaporates from the Earth's surface, rises and cools as it moves to higher elevations, condenses as rain or snow, and falls to the surface where it collects in lakes, oceans, and soil and in rocks underground.

9–12 Conservation of Energy and the Increase in Disorder

- Heat consists of random motion and the vibrations of atoms, molecules, and ions. The higher the temperature, the greater the atomic or molecular motion.

Related Ideas in *Benchmarks for Science Literacy* (AAAS 1993)

. . . .

K–2 The Earth

- Water left in an open container disappears, but water left in a closed container does not disappear.

3–5 The Earth

★ When liquid water disappears, it turns into a gas (vapor) in the air and can reappear as a liquid when cooled or as a solid if cooled below the freezing point of water. Clouds and fog are made of tiny droplets [or frozen

---

★ Indicates a strong match between the ideas elicited by the probe and a national standard's learning goal.

crystals] of water. (Note: The brackets indicate language added to the original benchmark. This revised benchmark appears in AAAS 2007, p. 21.)

- Air is a substance that surrounds us, takes up space, and whose movement we feel as wind.

6–8 The Earth

- Water evaporates from the surface of the Earth, rises and cools, condenses into rain or snow, and falls again to the surface.

6–8 Structure of Matter

- Atoms and molecules are perpetually in motion. Increased temperature means greater average energy of motion, so most substances expand when heated. In solids, the atoms are closely locked in position and can only vibrate. In liquids, the atoms or molecules have higher energy, are more loosely connected, and can slide past one another; some molecules may get enough energy to escape into a gas. In gases, the atoms or molecules have still more energy and are free of one another except during occasional collisions.

9–12 Structure of Matter

- An enormous variety of biological, chemical, and physical phenomena can be explained by changes in the arrangement and motion of atoms and molecules.

Related Research

- Before students can explain the process of condensation, they need to know where

the water that condenses comes from. Research has shown that students seem to transit a series of stages in their understanding of evaporation. At first, they may seem to think that when water evaporates, it simply ceases to exist. In the next stage, they may think it changes location but that it changes into some other form we can perceive, such as fog, steam, or droplets. Fifth grade is about the time that students can accept air as the location of evaporating water, if they have had special instruction that targets this idea (AAAS 1993).

- In a study of Israeli children ages 10–14 (Bar and Travis 1991), children were asked what causes water to form on the outside of a container holding ice. The most frequent responses were "the coldness changed into water" or "the cold caused hydrogen and oxygen to change into water." The researchers concluded that even though students knew that water as a gas (vapor) could be changed back to a liquid, applying that knowledge was difficult for students (Driver et al. 1994).

- In Osborne and Cosgrove's (1983) study of New Zealand children's ideas about change in state, they found that the percentage of students who could explain condensation as resulting from water in the air increased from 10% in children below age 12 to 55% in children ages 12–17. The majority of students gave macroscopic descriptions of condensation with very few mentioning energy and movement of molecules (Driver et al. 1994).

Suggestions for Instruction and Assessment

- Elementary school students need concrete experiences to understand what happens to water during a change in state before developing the more sophisticated idea of a water cycle.

- Observing this or similar phenomena and posing the question "Where do you think the water came from?" should be a part of every elementary student's science experience. For students who strongly believe the water came from within the container of ice cubes, confront their ideas with a sealed container of ice in which they agree that nothing can get in or out. Another way to confront them with their idea that the water somehow leaked out of the container is to freeze ice that has been colored with food coloring. As the ice melts inside the container and a pool of colored water collects inside the container, the condensation outside the container will be clear. Use this discrepant event to challenge their ideas and help them think through a new explanation.

- Do not assume that because students use words like *evaporation* and *condensation* they actually know what is happening. Develop the concept before giving students the technical term for the processes that make up the water cycle.

- Be aware that textbook and internet representations of the water cycle may contribute to the idea that evaporated water immediately goes up to the clouds or the Sun, rather than exists in the air around us.

Many representations show upward arrows pointing to the clouds or Sun that may influence students' thinking about condensation so that they do not recognize that condensed water on an object forms from the surrounding air that holds water vapor.

- Teaching about condensation in the water cycle involves several interrelated ideas that should be combined in instruction. These ideas include conservation of matter, phase change, and composition and nature of air. It is particularly important that students accept the idea of air being a substance that is made up of matter (including water) in a form we cannot see.

- Have students visually observe condensation. Place a container in a shallow dish of water. Place clear plastic wrap loosely over the dish and container with the edges sealed. Put a coin or some small weight in the middle of the plastic so it sags into the container and place the device in the sun. As the water warms, condensation will occur on the inside of the plastic and run down into the container. Have students draw diagrams to trace and explain how the water moved from the shallow dish to the sides of the plastic and into the container.

Related NSTA Science Store Publications and NSTA Journal Articles

American Association for the Advancement of Science (AAAS). 1993. *Benchmarks for science literacy.* New York: Oxford University Press.

American Association for the Advancement of Sci-

ence (AAAS). 2007. *Atlas of science literacy.* Vol. 2, "weather and climate," 20–21. Washington, DC: AAAS.

Driver, R., A. Squires, P. Rushworth, and V. Wood-Robinson. 1994. *Making sense of secondary science: Research into children's ideas.* London and New York: RoutledgeFalmer.

Gilbert, S. W., and S. W. Ireton. 2003. *Understanding models in Earth and space science.* Arlington, VA: NSTA Press.

Keeley, P. 2005. *Science curriculum topic study: Bridging the gap between standards and practice.* Thousand Oaks, CA: Corwin Press.

National Research Council (NRC). 1996. *National science education standards.* Washington, DC: National Academy Press.

Robertson, W. 2005. *Air, water, and weather: Stop Faking It! Finally Understanding Science So You Can Teach It.* Arlington, VA: NSTA Press.

Smith, M., and J. Southard. 2002. Water is all around you. *Science Scope* 26 (2): 32–35.

Related Curriculum Topic Study Guides

(Keeley 2005)

"States of Matter"

"Water Cycle"

References

American Association for the Advancement of Science (AAAS). 1993. *Benchmarks for science literacy.* New York: Oxford University Press.

Bar, V., and A. Travis. 1991. Children's views concerning phase changes. *Journal of Research in Science Teaching* 28 (4): 363–382.

Driver, R., A. Squires, P. Rushworth, and V. Wood-Robinson. 1994. *Making sense of secondary science: Research into children's ideas.* London and New York: RoutledgeFalmer.

Keeley, P. 2005. *Science curriculum topic study: Bridging the gap between standards and practice.* Thousand Oaks, CA: Corwin Press.

Keeley, P., F. Eberle, and L. Farrin. 2005. *Uncovering student ideas in science: 25 formative assessment probes.* Vol. 1. Arlington, VA: NSTA Press.

National Research Council (NRC). 1996. *National science education standards.* Washington, DC: National Academy Press.

Osborne, J., and M. Cosgrove. 1983. Children's conceptions of the changes of state of water. *Journal of Research in Science Teaching* 20 (9): 825–838.

Rainfall

A group of friends was looking up at a rain cloud. Each had a different idea about how rain falls from the clouds. This is what they said:

Will: "I think rain falls when clouds melt."

Chandra: "I think rain falls when the clouds open up."

Bettina: "I think rain falls when the clouds get too cold."

Fern: "I think rain falls when clouds are shaken or pushed together."

Lorna: "I think rain falls when the evaporated water in clouds condenses."

Marcus: "I think rain falls when water drops in the clouds get too heavy."

Jonah: "I think rain falls as the ice crystals in the clouds begin to melt."

Nona: "I think rain falls when the water falls through little holes in the clouds."

Which friend do you most agree with? _____

Explain your thinking about what causes rain to fall from the clouds.

_____

_____

_____

_____

_____

_____

Rainfall

Teacher Notes

Purpose

The purpose of this assessment probe is to elicit students' ideas about precipitation. The probe is designed to determine whether students understand what causes the water in clouds to fall as rain.

Related Concepts

clouds, gravity, precipitation, rain, water cycle

Explanation

The best answer is Marcus's: I think rain falls when water drops in the clouds get too heavy. Rain is the liquid form of precipitation. Precipitation is a complex process. A simple explanation begins with water vapor in the warm air rising in the sky, cooling to the dew point (where condensation occurs), and forming tiny drops of suspended liquid water. When there

are enough of these tiny drops of suspended liquid water, they accumulate to form clouds. As the drops accumulate, some of the drops will combine and form larger drops, and some drops will acquire more water vapor from the air. Clouds often appear gray when they contain larger water drops. Eventually the large drops are too heavy to remain suspended in the sky and succumb to the pull of gravity from the Earth. This pull of gravity causes the water to fall from the clouds toward Earth, resulting in rain.

Curricular and Instructional Considerations

Elementary Students

In the early elementary school grades, the emphasis should be on observing and describing weather, including rain events. Observing rain and recording data, such as observations of clouds when it rains, amount of rain, days when it rains, and temperature, helps students form a foundation for understanding weather phenomena. Pondering questions, such as why rain does not always fall from the clouds, helps students begin to understand that there are certain conditions necessary for rain to occur. At the upper elementary level, students are beginning to develop a basic conception of gravity as a pull toward the Earth. Students can begin to link the concept of rain falling toward the Earth with gravity, but an understanding of what is happening within the cloud that causes the rain to fall should wait until middle school.

Middle School Students

Middle school students expand on their elementary experiences in observing and describing rain clouds to more conceptual ideas about the water cycle, including the composition and formation of rain clouds and the mechanism of precipitation. Students' growing understanding of the relationship between the mass of an object, the upward force of rising air, and the downward force of gravity will help them account for why large drops of water fall toward the Earth while tiny drops stay suspended in clouds.

High School Students

At the high school level, students use their knowledge of physics- and chemistry-related concepts to explain weather phenomena in

the Earth system, such as rain. Their understanding of the water cycle is set into the larger context of matter cycling through Earth as a system. Recognizing the role of gravity is important to their growing understanding of Earth as a system because, without gravity, there would be no water cycle. Their deepening knowledge of heat, temperature, change in state, evaporation, condensation, and the force of gravity is helpful in understanding *why* the water cycle occurs and knowing the processes that make up the water cycle.

Administering the Probe

All students have seen rain fall from clouds in the sky, although in some geographic areas, rain is a more common occurrence. If possible, take students outside to view rain as it falls. This probe can be used with other probes in this book, such as "What Are Clouds Made Of?" (p. 155) and "Where Did the Water Come From?" (p. 163), or combined with "Wet Jeans" from Volume 1 of this series (Keeley, Eberle, and Farrin 2005) to create a cluster of water cycle–related probes.

Related Ideas in *National Science Education Standards* (NRC 1996)

K–4 Changes in Earth and Sky

- Weather changes from day to day and over the seasons. Weather can be described by measurable quantities such as temperature, wind direction and speed, and precipitation.

★ Indicates a strong match between the ideas elicited by the probe and a national standard's learning goal.

5–8 Structure of the Earth System

★ Water evaporates from Earth's surface, rises and cools as it moves to higher elevations, condenses as rain or snow, and falls back to the surface, where it collects in lakes, oceans, and soil and in rocks underground.

• Clouds, formed by the condensation of water vapor, affect weather and climate.

9–12 Motion and Forces

• Gravitation is the universal force that each mass exerts on any other mass.

Related Ideas in *Benchmarks for Science Literacy* (AAAS 1993)

. . . .

K–2 The Earth

• Some events in nature have a repeating pattern. The weather changes from day to day, but things such as temperature and the amount of rain (or snow) tend to be high, low, or medium in the same months of the year.

3–5 The Earth

• When liquid water disappears, it turns into a gas (vapor) in the air and can reappear as a liquid when cooled or as a solid if cooled below the freezing point of water. Clouds and fog are made of tiny droplets [or frozen crystals] of water. (Note: The brackets indicate language added to the original benchmark. This revised benchmark appears in AAAS 2007, p. 21.)

★ Things on or near Earth are pulled toward it by Earth's gravity.

6–8 The Earth

★ Water evaporates from the surface of the Earth, rises and cools, condenses into rain or snow, and falls again to the surface.

★ Everything on or anywhere near Earth is pulled toward Earth's center by gravitational force.

9–12 The Earth

★ The action of gravitational force on regions of different densities causes them to rise or fall.

Related Research

• Some younger students believe that clouds get scrambled and melt and rain occurs when clouds are shaken (Philips 1991).

• A study by Bar (1986) revealed that when students ages 5–7 were asked what causes rain, there was little evidence of a relationship between clouds and rain. Those who did see a link between rain and clouds often described clouds as bags of water kept high in the sky, and when they collide, they rip open and the water falls out. The same study found that at ages 6–9, students think the clouds open up to make rain. Building on that idea, students in grades 7–10 further described rain as the drops that fall through little holes in the clouds. At ages 9–10, some students think rain falls when clouds become cold or heavy (Driver et al. 1994).

★ Indicates a strong match between the ideas elicited by the probe and a national standard's learning goal.

- Field tests of this probe with fourth through eighth graders reveal that many students think that clouds are made up of water vapor and that, when the water vapor condenses, it falls as rain. These students did not recognize the clouds as being made up of tiny drops of water. They explain water as evaporating and going up to the clouds, then condensing and falling as rain, but fail to recognize that clouds can exist as tiny droplets of water long before it rains.

Suggestions for Instruction and Assessment

- Provide younger students with opportunities to make relatively long-term sky observations of clouds to begin to understand that rain does not occur just because clouds are present. Visually observing the changes that happen to clouds before it rains sets the stage for later explanations of what happens in clouds that results in precipitation.

- Do not assume that because students use words like *evaporation, condensation,* and *precipitation* they actually know what is happening. Develop the concept before giving students the technical term for the processes that make up the water cycle.

- Ask students to draw a sequence of pictures to show and explain what happens to the water drops that make up clouds before and during a rain.

- Be aware of poor diagrams of the water cycle that often show water evaporating and rising to form a white cloud and then moving to a dark cloud that appears to open up and spill out streams of water. Poor representations like this are pervasive on the internet and should be used with caution.

- When teaching about Earth's gravity as a force that affects all objects on or near the Earth, explicitly connect the fall of rain to gravity and the concept of mass. Challenge students to think about why the effect of gravity on a large water drop in a cloud is different from the effect of gravity on a tiny water drop or rising water molecules.

- Probe further to find out what students think clouds are made of by using the probe "What Are Clouds Made Of?" (p. 155). Some students think that when water evaporates, it forms a cloud made up of evaporated water and that rain is the result of the water condensing and falling. Understanding what clouds are made of is connected to students' ideas about the fall of rain.

Related NSTA Science Store Publications and NSTA Journal Articles

American Association for the Advancement of Science (AAAS). 1993. *Benchmarks for science literacy.* New York: Oxford University Press.

American Association for the Advancement of Science (AAAS). 2007. *Atlas of science literacy.* Vol. 2, "weather and climate," 20–21. Washington, DC: AAAS.

Crane. P. 2004. On observing the weather. *Science and Children* 41 (8): 32–36.

Driver, R., A. Squires, P. Rushworth, and V. Wood-Robinson. 1994. *Making sense of secondary sci-*

ence: Research into children's ideas. London and New York: RoutledgeFalmer.

Keeley, P. 2005. *Science curriculum topic study: Bridging the gap between standards and practice.* Thousand Oaks, CA: Corwin Press.

National Research Council (NRC). 1996. *National science education standards.* Washington, DC: National Academy Press.

Robertson, W. 2005. *Air, water, and weather: Stop Faking It! Finally Understanding Science So You Can Teach It.* Arlington, VA: NSTA Press.

Vowell, J., and M. Phillips. A drop through time. *Science and Children* 44 (9): 30–34.

Williams, J. 1997. *The weather book: An easy-to-understand guide to the USA's weather.* 2nd ed. Arlington, VA: NSTA Press.

Related Curriculum Topic Study Guides

(Keeley 2005)

"Air and Atmosphere"

"Weather and Climate"

References

American Association for the Advancement of Science (AAAS). 1993. *Benchmarks for science literacy.* New York: Oxford University Press.

American Association for the Advancement of Science (AAAS). 2007. *Atlas of science literacy.* Vol. 2, "weather and climate," 20–21. Washington, DC: AAAS.

Bar, V. 1986. *The development of the conception of evaporation.* Jerusalem, Israel: The Amos de-Shalit Science Teaching Center in Israel, The Hebrew University of Jerusalem, Israel.

Driver, R., A. Squires, P. Rushworth, and V. Wood-Robinson. 1994. *Making sense of secondary science: Research into children's ideas.* London and New York: RoutledgeFalmer

Keeley, P. 2005. *Science curriculum topic study: Bridging the gap between standards and practice.* Thousand Oaks, CA: Corwin Press.

Keeley, P., F. Eberle, and L. Farrin. 2005. *Uncovering student ideas in science: 25 formative assessment probes.* Vol. 1. Arlington, VA: NSTA Press.

National Research Council (NRC). 1996. *National science education standards.* Washington, DC: National Academy Press.

Philips, W. C. 1991. Earth science misconceptions. *The Science Teacher* 58 (2): 21–23.

Summer Talk

Six friends were talking. They each had different ideas about why it is warmer in the summer than in the winter. This is what they said:

Werner: "It's because the winter clouds block heat from the Sun."

Ava: "It's because the Sun gives off more heat in the summer than in winter."

Raul: "It's because Earth's tilt changes the angle of sunlight hitting Earth."

Fernando: "It's because the Earth orbits closer to the Sun in the summer than in the winter."

Shakira: "It's because one side of Earth faces the Sun and the other side faces away."

Susan: "It's because the Northern Hemisphere is closer to the Sun in summer than in the winter."

Which friend do you most agree with? _____

Describe your thinking about why it is warmer in the summer than in the winter. Provide an explanation for your answer.

_____

_____

_____

_____

Summer Talk

Teacher Notes

Purpose

The purpose of this assessment probe is to elicit students' ideas about seasons. The probe can be used to determine whether students recognize the effect of the Earth's tilt on its axis and the resulting intensity of sunlight as the reason for seasons.

Related Concepts

Earth-Sun system, seasons

Explanation

The best response is Raul's: It's because Earth's tilt changes the angle of sunlight hitting Earth. Seasons are primarily caused by the 23.5-degree angled tilt of Earth's axis as it revolves around the Sun in a slightly elliptical orbit (almost circular). This 23.5-degree angle varies slightly over time between 22.2 and 24.5 de-

grees. As Earth revolves around the Sun, this tilted axis always points in the same direction. This means that during part of the year one hemisphere will be leaning or bending more away from the Sun, which results in winter, and the other hemisphere will be leaning or bending more toward the Sun, which results in summer.

What does this have to do with summer being warmer? The tilt affects the intensity of sunlight striking Earth in different locations. When a hemisphere is tilted away from the Sun in the winter, the rays from the Sun strike this part of Earth at a lower angle that spreads the sunlight over a larger surface area. Therefore, some regions receive less heat, such as the Northern Hemisphere during winter. When it is winter in the Northern Hemisphere, the Southern Hemisphere is experiencing summer

because the Southern Hemisphere is tilted toward the Sun. The sun strikes Earth at a higher angle during the summer, which concentrates Earth's energy so it is less spread out. This increases the intensity of sunlight and thus warms the surface more. In the other half of Earth's orbit (half a year later) the situation reverses itself and it becomes summer in the Northern Hemisphere as the north pole is tilted toward the Sun and the south pole is tilted away, resulting in winter. In addition, during the summer, the Sun stays above the horizon longer, providing more time for the Sun's energy to heat that region of Earth.

Sunlight is most concentrated near the equator because of the more direct rays of the Sun. The number of hours of daylight and hours of darkness are almost the same. This unchanging angle of sunlight and consistent daytime and nighttime results in a year with minimal seasonal change.

Many people think the reason for the seasons is the Sun's closer proximity to Earth in the summer than in the winter. It is true that at times in Earth's slightly elliptical orbit we are closer to the Sun. However, we are actually closer in the winter than in the summer; thus distance is not a reason for why it is warmer in the summer. The Earth is closest to the Sun on January 4 and farthest away from the Sun in July, when we have our Fourth of July barbeques.

Curricular and Instructional Considerations

· ·

Elementary Students

In the elementary school grades, students learn about the four seasons and the changes that happen during each season. The focus is on observations, not explanations of what causes the seasons, an idea that is much too complex for this grade level.

Middle School Students

Students at this level begin to develop understandings of the Earth-Sun system, including how Earth orbits the Sun and the role of sunlight in heating Earth. However, explaining what causes the seasons is still very difficult at this age because of the complex spatial reasoning required to understand the seasons. Nevertheless, an explanation is introduced at this grade level, though it may not be until high school that students can fully comprehend it.

High School Students

At this level, students' understanding of orbital geometry is more developed and enables them to better understand the reason why we have seasons. Their quantitative understanding of the intensity of light based on the area of light striking a surface helps them understand how the angle of sunlight reaching different parts of Earth results in the warming that produces seasonal variations. Because the idea of seasons is so difficult to learn at earlier grades, it is recommended that it be revisited in high school.

Administering the Probe

This probe is best used at the middle or high

school level to determine students' misconceptions about seasons before instruction.

Related Ideas in *National Science Education Standards* (NRC 1996)

K–4 Objects in the Sky

- The Sun provides the light and heat necessary to maintain the temperature of Earth.

K–4 Changes in the Earth and Sky

- Weather changes from day to day and over the seasons. Weather can be described by measurable quantities such as temperature, wind direction and speed, and precipitation.

5–8 Earth in the Solar System

- ★ The Sun is the major source of energy for phenomena on Earth's surface, such as growth of plants, winds, ocean currents, and the water cycle. Seasons result from variations in the amount of the Sun's energy hitting Earth's surface due to the tilt of Earth's rotation on its axis and the length of the day.

Related Ideas in *Benchmarks for Science Literacy* (AAAS 1993)

K–2 The Earth

- Some events in nature have a repeating pattern.

K–2 Energy Transformations

- The Sun warms the land, air, and water.

6–8 The Earth

- ★ Because Earth turns daily on an axis that is tilted relative to the plane of Earth's yearly orbit around the Sun, sunlight falls more intensely on different parts of Earth during the year. The variation in heating of Earth's surface produces the planet's seasons and weather patterns. (Note: See current revision of this benchmark below, under "9–12 The Earth.")

- ★ The temperature of a place on Earth's surface tends to rise and fall in a somewhat predictable pattern every day and over the course of a year. The pattern of temperature changes observed in a place tend to vary depending on how far north or south of the equator the place is, how near to oceans it is, and how high above sea level it is. (Note: This is a new benchmark. It can be found in AAAS 2007, p. 21.)

- ★ The number of hours of daylight and the intensity of the sunlight both vary in a predictable pattern that depends on how far north or south of the equator the place is. This variation explains why temperatures vary over the course of the year and at different locations. (Note: This is a new benchmark. It can be found in AAAS 2007, p. 21.)

6–8 Energy Transformations

- Light and other electromagnetic waves can warm objects. How much an object's tem-

---

★ Indicates a strong match between the ideas elicited by the probe and a national standard's learning goal.

perature increases depends on how intense the light striking its surface is, how long the light shines on the object, and how much of the light is absorbed. (Note: This is a new benchmark. It can be found in AAAS 2007, p. 21.)

9–12 The Earth

★ Because Earth turns daily on an axis that is tilted relative to the plane of Earth's yearly orbit around the Sun, sunlight falls more intensely on different parts of Earth during the year. The difference in intensity of sunlight and the resulting warming of Earth's surface produces the seasonal variations in temperature. (Note: This is a new benchmark for grades 9–12, revised from the original grades 6–8 benchmark. It can be found in AAAS 2007, p. 21.)

Related Research

- Students of all ages (including college students and adults) have difficulty understanding what causes the seasons (AAAS 2007).
- Studies by Sadler (1987) and Vosniadou (1991) show that students may not be able to understand an explanation of the seasons until they can reasonably understand the relative size, motion, and distance of the Sun and Earth (AAAS 2007).
- Atwood and Atwood (1996), Dove (1998), Philips (1991), and Sadler (1998) found that many students up through preservice believe that winter is colder than summer because the Earth is farther from the Sun

in the winter. This belief persisted even after instruction in Earth science (AAAS 2007).

- Researchers have attributed students' thinking that it is warmer in summer time because the Earth is closer to the Sun to their conception of an elongated elliptical orbit which makes it appear as if the Sun is closer to Earth during certain times of the year (Sadler 1998).
- Galili and Lavrik (1998) found that students' ideas about how light travels may interfere with their understanding of the seasons.
- Some students confuse Earth's daily rotation with its yearly revolution around the Sun (Salierno, Edleson, and Sherin 2005). For example, they believe that the side of Earth not facing the Sun experiences winter (AAAS 2007).
- Baxter (1989) noted age-related trends in children's alternative conceptions related to seasons. The most common conception in children up through the age of 16 was that the Sun is farther away in the winter. This was consistently prevalent across ages and particularly pervasive in the 9–12 age range. Up to age 10, a small number of students think changes in plants cause the seasons. This number increases between the ages of 10 and 12. A small number of students up to ages 9 and 10 think winter clouds stop the heat from the Sun. This idea increases slightly between ages 10 and 14. Up until age 12, some students think the Sun moves to the other side of Earth. By

---

★ Indicates a strong match between the ideas elicited by the probe and a national standard's learning goal.

age 12, this idea seems to go away (Driver et al. 1994).

Suggestions for Instruction and Assessment

- Students need to understand Earth's orientation in relation to the Sun and how sunlight strikes the Earth before they can develop explanations for the seasons.

- When teaching about the shape of Earth's orbit, the word *ellipse* can be confusing to most students and to some teachers. For many, *ellipse* means a highly oval-like orbit and they do not recognize a circle as being a special ellipse (just like a square is a special type of rectangle). Once they realize a circle as a special ellipse, they are more accepting of ellipses that are almost indistinguishable from circles.

- Be aware that textbook representations that show an exaggerated elliptical orbit can reinforce the idea that the Earth is closer to the Sun during parts of the year. Explicitly point out these flaws to students and show why an exaggerated ellipse is used in representations. An analogy you can use is to hold a glass or cup so you are looking straight down into the glass and the rim will look like a circle. Next hold the glass or cup out in front of you slightly below the level of your eyes. Now when you look at the rim it looks like an exaggerated ellipse. The rim did not change. The only thing that changes was your viewing plane. Relate this to the view in textbook representations.

- The PRISMS website, a National Science Foundation–funded National Science Digital Library project, provides reviews of representations and phenomena used to teach the seasons. These can be accessed at *http://prisms.mmsa.org*.

- Select carefully designed curricula to teach the reasons for seasons. The Great Explorations in Math and Science (GEMS) unit *The Real Reasons for Seasons* and Project ASTRO have lessons developed to address commonly held ideas (Gould, Willard, and Pompea 2000; Fraknoi 1995).

- Use models to show how the angle of sunlight affects how much surface area a given amount of light covers. Relate light energy to heat, demonstrating how the same amount of heat in a less spread-out area warms a surface more than the same amount of heat that is more spread out.

- Obtain images of the Sun taken at different times of the year from the same location and camera position (see NASA Sun-Earth resources at *http://sunearth.gsfc.nasa.gov*). Enlarge the images using the same percentage of enlargement. Students can use the enlarged images to measure the Sun's diameter and make the link between increased diameter and closer distance to Earth. Students will find that the diameter of the Sun in winter is slightly more than in summer, thus dispelling their notion that the Sun is closer to Earth in the summer. This contradiction of their belief that the Sun is closer in the summer encourages students to think and seek out what a pos-

sible explanation could be.

- *Kinesthetic Astronomy,* from the Space Science Institute in Boulder, Colorado, is a good resource for modeling seasons. It can be downloaded free of charge at *www.spacescience.org/education/extra/kinesthetic_astronomy/index.html.*

Related NSTA Science Store Publications and Journal Articles

American Association for the Advancement of Science (AAAS). 1993. *Benchmarks for science literacy.* New York: Oxford University Press.

American Association for the Advancement of Science (AAAS). 2007. *Atlas of science literacy.* Vol. 2. "weather and climate," 20–21. Washington, DC: AAAS.

Driver, R., A. Squires, P. Rushworth, and V. Wood-Robinson. 1994. *Making sense of secondary science: Research into children's ideas.* London and New York: RoutledgeFalmer.

Gilbert, S., and S. Ireton. 2003. *Understanding models in Earth and space science.* Arlington, VA: NSTA Press.

Keeley, P. 2005. *Science curriculum topic study: Bridging the gap between standards and practice.* Thousand Oaks, CA: Corwin Press.

National Research Council (NRC). 1996. *National science education standards.* Washington, DC: National Academy Press.

Philips, W. C. 1991. Earth science misconceptions. *Science Teacher* 58 (2): 21–23.

Plait, P. 2002. *Bad astronomy: Misconceptions and misuses revealed.* New York: John Wiley and Sons.

Robertson, W. 2007. Science 101: What causes the seasons? *Science and Children* (Dec.): 54–57.

Related Curriculum Topic Study Guide

(Keeley 2005)
"Seasons"

References

American Association for the Advancement of Science (AAAS). 1993. *Benchmarks for science literacy.* New York: Oxford University Press.

American Association for the Advancement of Science (AAAS). 2007. *Atlas of science literacy.* Vol. 2, "weather and climate," 20–21. Washington, DC: AAAS.

Atwood, R., and V. Atwood. 1996. Preservice elementary teachers' conceptions of the causes of seasons. *Journal of Research in Science Teaching* 33: 553–563.

Baxter, J. 1989. Children's understanding of familiar astronomical events. *International Journal of science Education* 11 (special issue): 502–513.

Dove, J. 1998. Alternative conceptions about the weather. *School Science Review* 79: 65–69.

Driver, R., A. Squires, P. Rushworth, and V. Wood-Robinson. 1994. *Making sense of secondary science: Research into children's ideas.* London and New York: RoutledgeFalmer.

Fraknoi, A., ed. 1995. *The universe at your fingertips.* San Francisco: Astronomical Society of the Pacific.

Galili, I., and V. Lavrik. 1998. Flux concept in learning about light: A critique of the present situation. *International Journal of Science Edu-*

cation 82: 591–613.

Gould, A., C. Willard, and S. Pompea. 2000. *The real reason for seasons—Sun-Earth connection.* Berkeley, CA: Lawrence Hall of Science.

Keeley, P. 2005. *Science curriculum topic study: Bridging the gap between standards and practice.* Thousand Oaks, CA: Corwin Press.

National Research Council (NRC). 1996. *National science education standards.* Washington, DC: National Academy Press.

Philips, W. C. 1991. Earth science misconceptions. *Science Teacher* 58 (2): 21–23.

Sadler, P. 1987. Misconceptions in astronomy. In *Proceedings of the second international seminar on misconceptions and educational strategies in science and mathematics.* Vol. 3. Ed. J. Novak,

422–425. Ithaca, NY: Cornell University.

Sadler, P. 1998. Psychometric models of student conceptions in science: Reconciling qualitative studies and distracter-driven assessment items. *Journal of Research in Science Teaching* 35: 265–296.

Salierno, C., D. Edleson, and B. Sherin. 2005. The development of student conceptions of the Earth-Sun relationship in an inquiry-based curriculum. *Journal of Geoscience Education* 53: 422–431.

Vosniadou, S. 1991. Designing curricula for conceptual restructuring: Lessons from the study of knowledge acquisition in astronomy. *Journal of Curriculum Studies* 23: 219–237.

Me and My Shadow

Five friends were looking at their shadows early one morning. They wondered what their shadows would look like by the end of the day. This is what they said:

Jamal: "My shadow will keep getting longer throughout the day."

Morrie: "My shadow will keep getting shorter throughout the day."

Amy: "My shadow will keep getting longer until it reaches its longest point and then it will start getting shorter."

Fabian: "My shadow will keep getting shorter until noon and then it will start getting longer."

Penelope: "My shadow will stay about the same from morning to day's end.

Which friend do you most agree with? _____

Describe your thinking. Explain the reason for your answer.

_____

_____

_____

_____

_____

_____

_____

_____

Me and My Shadow

Teacher Notes

Purpose

The purpose of this assessment probe is to elicit students' ideas about light and shadows. The probe is designed to find out students' ideas about how shadows change throughout the day.

Related Concepts

Earth-Sun system, shadows

Explanation

The best response is Fabian's: My shadow will keep getting shorter until noon and then it will start getting longer. Shadows are longest right after sunrise and right before sunset. The angle at which sunlight strikes Earth's surface changes as the Sun appears to move across the sky due to the rotation of Earth. In the early morning, the Sun is low on the horizon and the angle be-

tween the Sun and Earth's surface is small. The shadow that results from blocking the Sun's rays is long. As the angle between Earth's surface and the Sun's rays that strike Earth's surface increases throughout the morning, the shadow gets smaller. Noon is defined as the time when the Sun is highest in the sky. At noon, the size of a shadow is shortest and will begin increasing. Throughout the afternoon, the shadow gradually grows longer and its position is now on the other side of the object. It reaches its longest length just before sundown when the angle between the Sun's rays and Earth's surface is small and the Sun is once again low on the horizon. The sequence of shadow length is such that it starts out very long on one side of the object, gradually shortens and then gradually lengthens on the opposite side of the object until night, when there is no more sunlight to cast a shadow.

Curricular and Instructional Considerations

Elementary Students

Observing changes in shadows is a common activity for elementary school students. It is a good way to build inquiry skills and identify patterns. In the early grades, the changing length of shadows is primarily observational. In the later elementary grades, students begin to relate the changing size and position of the shadow to the position of the Sun in the sky in relation to Earth's surface in order to explain how shadows change.

Middle School Students

Middle school students can extend their investigations of shadows to look for seasonal patterns. At this level, students understand that the position of the daytime Sun is related to Earth's rotation and results in the length and position of shadows. Students also develop the notion that the shortest shadow on any given day always points due north and that the length of the shadow at noon varies with the seasons. Using models, middle school and high school students can investigate how variations in length of a shadow at noon relates to Earth's 23.5-degree tilt as it revolves around the Sun. This also helps build students' understanding of why we have seasons and why seasons in the Northern and Southern Hemispheres are opposite. Students can combine their knowledge of shadow formation with technological design by constructing sundials or examining how ancient cultures used shadows to mark certain days of the year. Students explore other ideas about shadows from light sources other than the Sun.

High School Students

At this level, students use ideas about light reflection to further investigate how shadows are formed, including comparing shadows formed by point sources versus extended sources of light. Students make quantitative comparisons between the size of a shadow and the distance of an object from a light source. However, even though students' shadows on a sunny day are a familiar phenomenon, some high school students still have difficulty picturing how a shadow changes throughout the day.

Administering the Probe

Make sure students understand that the probe is asking about what happens to the length of the shadow from the time the Sun rises to the time the Sun sets. Consider having students draw a sequence of pictures showing the shadow in relation to the Sun during different parts of the day.

Related Ideas in *National Science Education Standards* (NRC 1996)

K–4 Objects in the Sky

★ The Sun, Moon, stars, clouds, birds, and airplanes all have properties, locations, and movements that can be observed and described.

★ Indicates a strong match between the ideas elicited by the probe and a national standard's learning goal.

K–4 Changes in the Earth and Sky

★ Objects in the sky have patterns of movement. The Sun, for example, appears to move across the sky in the same way every day, but its path changes slowly over the seasons.

K–4 Light, Heat, Electricity, and Magnetism

• Light travels in a straight line until it strikes an object.

5–8 Transfer of Energy

• Light interacts with matter by transmission (including refraction), absorption, or scattering (including reflection).

5–8 Earth in the Solar System

• Most objects in the solar system are in regular and predictable motion.

Related Ideas in *Benchmarks for Science Literacy* (AAAS 1993)

K–2 The Universe

• The Sun, Moon, and stars all appear to move slowly across the sky.

K–2 The Earth

★ Some events in nature have a repeating pattern.

3–5 Motion

• Light travels and tends to maintain its direction of motion until it interacts with an ob-

ject or material. (Note: This is a new benchmark. It can be found in AAAS 2001.)

3–5 Constancy and Change

★ Things change in steady, repetitive, or irregular ways—or, sometimes, in more than one way at the same time. Often the best way to tell which kinds of change are happening is to make a table or graph of measurements.

Related Research

• Some researchers have found that children expect the shadow of an object to be the same shape as the object. Their predictions about shadows often refer to a shadow as a "reflection" or as a "dark reflection" on a screen (Driver et al. 1994).

• Students seem to have more success in locating where an object's shadow will fall in relation to a light source if the object is a person. They have more difficulty anticipating where a shadow will fall if it is a nonhuman object, such as a tree (Driver et al. 1994).

• Students who were able to anticipate where a shadow would fall and explain their ideas in terms of relative position of the light source, object, and the object's shadow did so without including an explanation of the straight path of light in their explanation (Driver et al. 1994).

Suggestions for Instruction and Assessment

• This probe lends itself to an inquiry investigation. After students commit to an idea

★ Indicates a strong match between the ideas elicited by the probe and a national standard's learning goal.

on the probe, have them test their idea by going outside throughout the day and observing the lengths of their shadows. Students should be asked to make predictions about how the length and direction of the shadow will change over the next hour or so and be able to test their predictions.

- Make sure students understand what a shadow is before asking them to explain how shadows change.

- Have students model what happens to a shadow when the position of a light source, and thus the angle at which light strikes an object, changes. Provide students with flashlights and an upright object to test their ideas and record observations. Help students link their flashlight findings to the position of the Sun throughout the day.

- Consider having students draw a sequence of pictures, like a storyboard, to show the relationship between a shadow's length and position and the position of the Sun throughout the day.

- Extend shadow investigations across the year to show that there are seasonal patterns as well as daily patterns.

- Have students investigate how ancient cultures used their knowledge of shadows to mark seasonal events and tell time.

- Consider having students participate in NSTA's "Astronomy with a Stick" project. Information on this project can be found at *www3.nsta.org/awsday*.

- Have students use mathematics to figure out how high the Sun is at noon during each of the four seasons to show that the height varies throughout the year, which accounts for why shadow size at noon varies throughout the year. On the first day of spring and fall, the Sun is 90 degrees minus your latitude above the horizon at noon (for example, 90 degrees – 40 degrees = 50 degrees for New Jersey). For noon on the first day of summer, add 23.5 degrees (the angle of Earth's tilt) to that (50 degrees + 23.5 = 73.5 degrees for New Jersey) and for noon on the first day of winter subtract 23.5 degrees from that (50 degrees – 23.5 = 26.5 degrees for New Jersey). Have students try this for their locations and compare the measurements of their shortest shadow during the first day of fall, winter, spring, and summer. Students may be quite surprised to see that in some locations, such as New Jersey, the noon shadow on the first day of winter can be quite long.

Related NSTA Science Store Publications and NSTA Journal Articles

American Association for the Advancement of Science (AAAS). 1993. *Benchmarks for science literacy.* New York: Oxford University Press.

American Association for the Advancement of Science (AAAS). 2001. *Atlas of science literacy.* Vol. 1, "changes in the Earth's surface," 50–51. Washington, DC: AAAS.

Barrows, L. 2007. Bringing light onto shadows. *Science and Children* 44 (9): 43–45.

Keeley, P. 2005. *Science curriculum topic study: Bridging the gap between standards and practice.* Thousand Oaks, CA: Corwin Press.

National Research Council (NRC). 1996. *National science education standards.* Washington, DC: National Academy Press.

Related Curriculum Topic Study Guides

(Keeley 2005)

"Earth, Moon, and Sun System"

"Visible Light, Color and Vision"

References

American Association for the Advancement of Science (AAAS). 1993. *Benchmarks for science literacy.* New York: Oxford University Press.

American Association for the Advancement of Science (AAAS). 2001. *Atlas of science literacy.* Vol. 1, "changes in the Earth's surface," 50–51. Washington, DC: AAAS.

Driver, R., A. Squires, P. Rushworth, and V. Wood-Robinson. 1994. *Making sense of secondary science: Research into children's ideas.* London and New York: RoutledgeFalmer.

Keeley, P. 2005. *Science curriculum topic study: Bridging the gap between standards and practice.* Thousand Oaks, CA: Corwin Press.

National Research Council (NRC). 1996. *National science education standards.* Washington, DC: National Academy Press.

Where Do Stars Go?

Five friends were wondering where stars were in the daytime. They each had different ideas about why we do not see stars in the sky during the day. This is what they said:

Jack: "The stars stop shining when the Sun comes out."

Shelley: "The stars are still in the sky above us, but we can't see them."

Nancy: "The stars go underneath Earth during the daytime."

Emma: "The stars cool down during the day and the Sun gets hotter."

Flavio: "The stars are on the other side of Earth where it's nighttime."

Which friend do you most agree with? _____

Describe your thinking about why you do not see stars during the daytime. Provide an explanation for your answer.

_____

_____

_____

_____

_____

_____

_____

_____

_____

_____

Where Do Stars Go?

Teacher Notes

Purpose

The purpose of this assessment probe is to elicit students' ideas about stars. The probe is designed to examine students' ideas about the location of stars in the daytime.

Related Concept

stars

Explanation

The best response is Shelley's: The stars are still in the sky above us, but we cannot see them. The reason we see stars at night is because our location on Earth is turned away from the Sun, blocking its light so that the evening sky is dark. During the day, the bright light from our Sun, our closest star, illuminates the sky overhead and prevents our eyes from seeing the fainter light from distant stars. We can only see them in the darkness of the night sky. This also explains why we only see some stars as night falls. The stars that appear brightest to us become visible first and gradually other stars appear as the night sky darkens.

The location of the stars and the patterns they make in our sky do not change. In the daytime, they are still in the overhead sky in the same pattern as we would see it if it were night. However, their positions in the overhead sky change because of the daily rotation of Earth, which makes it appear as if the stars and their patterns are moving across the sky from east to west. As Earth revolves around the Sun over the course of a year, different star patterns appear at night as we look in the direction away from the Sun and others become invisible as they appear in the direction of the Sun. This is similar to riding a merry-go-round as we face out to see our parents or friends who are standing around and watching us. This explains why we see different stars at different seasons. Without Earth's revolution, a given star or star pattern would appear to rise at the same time each day; however, because of Earth's revolution they appear to rise about four minutes earlier each day. There are some star patterns that we can see throughout the year because they are near the Pole Star (Polaris). These stars never set and are called *circumpolar stars.*

Besides the fixed stars and the patterns they make, there are a few bright lights in the sky that change their positions with respect to the fixed star patterns. These are the planets. The word *planet* is derived from the Greek word *planetes,* which means "wanderers." The brightest of these planets (much brighter than any of the fixed stars) is Venus, which may appear as a very bright light a few hours before sunrise or a few hours after sunset. Rivaling Venus in brightness is the planet Jupiter, while the planet Saturn and the red-colored planet Mars are as bright as the brightest stars. The much fainter planet Mercury can only be seen a few days of the year, either directly after sunset or directly before sunrise.

Curricular and Instructional Considerations

Elementary Students

In the elementary school grades, students make observations of the day and night sky and question where some objects, like the

Moon and stars, go during the daytime. This is the time to help students connect their observations of light to why we see stars at night and not during the day. At this level, students investigate the patterns of stars and observe that, although star pattern positions relative to Earth may change in the sky over the seasons, the patterns of stars remain the same. Upper elementary students can also observe how the position of planets, including Earth, changes in relation to the pattern of stars.

Middle School Students

In middle school, students' understanding of stars can include the vast distances between Earth and the stars and galaxies and how we can see many more stars in the sky with the use of a telescope. At this level, students should be able to explain how the position of the stars changes as Earth rotates and seasons change during Earth's revolution around the Sun.

High School Students

At this level, students learn more complex ideas related to stars, such as evolution, composition, and behavior. However, be aware that some students may retain earlier developed misconceptions about stars.

Administering the Probe

This probe is best used at the elementary school level to determine students' misconceptions about stars before instruction. However, it is useful in finding out if middle school and high school students have retained the misconception about where stars are located in the daytime.

Related Ideas in *National Science Education Standards* (NRC 1996)

K–4 Objects in the Sky

★ The Sun, Moon, stars, clouds, birds, and airplanes all have properties, locations, and movements that can be observed and described.

K–4 Changes in the Earth and Sky

• Objects in the sky have patterns of movement.

5–8 Earth in the Solar System

• Most objects in the solar system are in regular and predictable motion.

Related Ideas in *Benchmarks for Science Literacy* (AAAS 1993)

K–2 The Universe

• There are more stars in the sky than anyone can easily count, but they are not scattered evenly, and they are not all the same in brightness or color.

• The Sun can be seen only in the daytime, but the Moon can be seen sometimes at night and sometimes during the day. The Sun, Moon, and stars all appear to move slowly across the sky.

3–5 The Universe

★ The patterns of stars in the sky stay the same, although they appear to move across the sky nightly, and different stars can be seen in different seasons.

• Planets change their positions against the background of stars.

• Stars are like the Sun, some being smaller and some larger, but so far away that they look like points of light.

6–8 The Universe

• The Sun is many thousands of times closer to Earth than any other star. Light from the Sun takes a few minutes to reach Earth, but light from the next nearest star takes a few years to arrive.

Related Research

• Young children's ideas about the stars may be related to their ideas about the shape of Earth and day and night. Some younger students conceptualize a flat Earth, with the Sun above us and stars below in the daytime and the opposite at night. Even students who have a concept of a spherical Earth may still think in terms of above and below rather than being surrounded by sky (Driver et al. 1994).

• Agan (2004) points out that little astronomy education research has been done around students' ideas related to stars and galaxies. This probe is useful in generating teacher research regarding K–8 students' ideas about the location of stars. Field-test results from this probe show that a significant number of elementary school students in grades 1–4 have alternative ideas about where stars are located during the

---

★ Indicates a strong match between the ideas elicited by the probe and a national standard's learning goal.

day that appear to be connected to their understanding of the position of Earth in relation to other objects as it completes its daily rotation. These alternative ideas appear to decrease between grades 3–4.

- Field-test results of this probe indicate that the most commonly chosen distracter at the early elementary school level is the idea that the stars are beneath us during the daytime. Students describe how the stars "set" by going "down" beneath Earth during the daytime in much the same way they believe the Sun sets in the evening by going down beneath Earth. An increasing number of upper elementary school students and middle school students select the distracter that the stars are on the opposite side of Earth during the daytime. They use reasoning similar to their ideas about the day-night cycle—that the light from the stars is not visible in the daytime because the stars are on the opposite side of Earth where it is night.

Suggestions for Instruction and Assessment

- When teaching elementary students about the day-night cycle, the emphasis is usually on why we see the Sun during the day and not at night. Be aware that developing the idea that the Sun is facing away from Earth during nighttime might lead to the opposite conception that the stars are facing away from Earth in the daytime.
- Be aware that the use of common words and phrases like *sundown* and *the sun sets*

might imply to younger students that the Sun goes beneath Earth at night and the stars come up. Conversely, children might think that when the Sun "rises," the stars go back down beneath Earth, like the Sun does during the night.

- Use models to illustrate how the pattern of stars during an evening stays the same but their location overhead, relative to where you are viewing them as Earth rotates during a 24-hour period, changes.
- The model described above can be combined with a demonstration of how the light from a flashlight seen during the day (simulated in a well-lit room) is much harder to see than the light from the same flashlight at night (lights off in the room). Connect this to not being able to see the stars in the daytime even though they are emitting light to the brightness of the light from the Sun.

Related NSTA Science Store Publications and Journal Articles

American Association for the Advancement of Science (AAAS). 1993. *Benchmarks for science literacy.* New York: Oxford University Press.

American Association for the Advancement of Science (AAAS). 2001. *Atlas of science literacy.* Vol. 2, "stars," 46–47. Washington, DC: AAAS.

Driver, R., A. Squires, P. Rushworth, and V. Wood-Robinson. 1994. *Making sense of secondary science: Research into children's ideas.* London and New York: RoutledgeFalmer.

Keeley, P. 2005. *Science curriculum topic study: Bridging the gap between standards and practice.* Thousand Oaks, CA: Corwin Press.

National Research Council (NRC). 1996. *National science education standards.* Washington, DC: National Academy Press.

Riddle, B. 2006. Scope on the skies: Location, location, location. *Science Scope* (Jan.): 60–62.

Related Curriculum Topic Study Guide

(Keeley 2005)
"Stars and Galaxies"

References

Agan, L. 2004. Stellar ideas: Exploring students' understanding of stars. *Astronomy Education Review* 3 (1): 77–97.

American Association for the Advancement of Science (AAAS). 1993. *Benchmarks for science literacy.* New York: Oxford University Press.

Driver, R., A. Squires, P. Rushworth, and V. Wood-Robinson. 1994. *Making sense of secondary science: Research into children's ideas.* London and New York: RoutledgeFalmer.

Keeley, P. 2005. *Science curriculum topic study: Bridging the gap between standards and practice.* Thousand Oaks, CA: Corwin Press.

National Research Council (NRC). 1996. *National science education standards.* Washington, DC: National Academy Press.

Index